The Lymphatic System

Other titles in
Human Body Systems

The Lymphatic System

Julie McDowell and Michael Windelspecht

HUMAN BODY SYSTEMS
Michael Windelspecht, Series Editor

Greenwood Press
Westport, Connecticut • London

Library of Congress Cataloguing-in-Publication Data is available at www.loc.gov

British Library Cataloguing in Publication Data is available.

ISBN: 0–313–32494–8

First published in 2004

Greenwood Press, 88 Post Road West, Westport, CT 06881
An imprint of Greenwood Publishing Group, Inc.
www.greenwood.com

Printed in the United States of America

The paper used in this book complies with the
Permanent Paper Standard issued by the National
Information Standards Organization (Z39.48–1984).

10 9 8 7 6 5 4 3 2 1

Illustrations, unless otherwise credited, are by Sandy Windelspecht.

The *Human Body Systems* series is a reference, not a medical or diagnostic manual. No portion of this series is intended to supplement or substitute medical attention and advice. Readers are advised to consult a physician before making decisions related to their diagnosis or treatment.

Contents

Color photos follow p. 36

Series Foreword

Human Body Systems is a ten-volume series that explores the physiology, history, and diseases of the major organ systems of humans. An organ system is defined as a group of organs that physiologically function together to conduct an activity for the body. In this series we identify ten major functions. These are listed in Table F.1, along with the name of the organ system responsible for the activity. It is sometimes difficult to specifically define an organ system, because many of our organs have dual functions. For example, the liver interacts with both circulatory and digestive systems, the hypothalamus acts as a junction between the nervous and endocrine systems, and the pancreas has both digestive and endocrine secretions. This complex interaction of organs and tissues in the human body is still not completely understood.

This series is unique in that it provides a one-stop reference source for anyone with an interest in the human body. Whereas other references frequently cover one aspect of human biology, from anatomy and physiology to the prevention of diseases, this series takes a more holistic approach. Each volume not only includes a physiological description of how the system works from the cellular level upward, but also a historical summary of how research on the system has changed since the time of the ancients. This is an important aspect of the series, and one that is frequently overlooked in modern textbooks. In order to understand the successes and problems of modern medicine, it is first important to recognize not only the achievements of the past but also the misunderstandings and challenges of the pioneers in medical research.

For example, a visit to any major educational institution reveals large lec-

TABLE F.I. Organ Systems of the Human Body

Organ System	General Function	Examples
Circulatory	Movement of chemicals through the body	Heart
Digestive	Supply of nutrients to the body	Stomach, small intestine
Endocrine	Maintenance of internal environmental conditions	Thyroid
Lymphatic	Immune system, transport, return of fluids	Spleen
Muscular	Movement	Cardiac muscle, skeletal muscle
Nervous	Processing of incoming stimuli and coordination activity	Brain, spinal cord
Reproductive	Production of offspring	Testes, ovaries
Respiratory	Gas exchange	Lungs
Skeletal	Support, storage of nutrients	Bones, ligaments
Urinary	Removal of waste products	Bladder, kidneys

ture halls, where science instructors present material to the students on the anatomy and physiology of the human body. Sometimes these classes include laboratory sessions, but in the study of human biology, especially for students who are not bound for professional schools in medicine, the student's exposure to human biology typically centers on a two-dimensional graphic. Most educators accept this process as a necessary evil of the educational system, but few recognize that, in fact, the large lecture classroom is the product of a change in Egyptian religious beliefs before the start of the current era. During the decline of the Egyptian empires and the simultaneous rise of the ancient Greek culture, the Egyptian religious organizations began to forbid the dissection of the human body. This had a twofold influence on medicine. First, the ending of human dissections meant that medical professionals required lectures from educators instead of participation in laboratory-based education, which led to the birth of the lecture hall. The second consequence would plague modern medicine for a thousand years. Stripped of their access to human cadavers, researchers studied other "lesser" animals and extrapolated their findings to humans. The practices of the ancient Greeks were passed on over the ages and became the basis for the study of modern medicine. These traditions continue to this day throughout the educational institutions of the world.

The history of human biology parallels the development of modern science. In the seventeenth century, William Harvey's study of blood circulation challenged the long-standing belief of the ancient Greeks that blood was produced in the liver and consumed in the tissues of the body. Harvey's pioneering experimental work had a strong influence on others, and within a century the legacy of the ancient Greeks had collapsed. In the eighteenth century a group of chemists who focused on the chemical reactions of the human body, called the iatrochemists, began to apply chemical laws to human physiology. They were joined by the iatrophysicists, who believed that the human body must operate under the physical laws of the universe. This in turn led to the beginnings of organic chemistry and biochemistry in the nineteenth century, as scientists focused on identifying the building blocks of living cells and the chemical reactions that they utilize in their metabolism.

In the past century, especially in the last three decades, the rapid advances in technology and scientific discovery have tended to separate most sciences from the general public. Yet despite an ongoing trend to leave the majority of the physical sciences to the scientists, interest in human biology has actually increased among the general population. This is primarily due to medical discoveries that increase not only lifespan but also healthspan, or the number of years that people live disease free. But another important aspect of this trend is the desire among the general public to be able to ask intelligent questions of their physicians and seek additional information on prescribed medications or procedures. In many cases this information serves as a system of checks and balances on the medical profession, ensuring that the patient is kept well informed and aware of the fundamentals regarding the procedure.

This is one of the most remarkable ages in the study of human biology. The recently announced completion of the Human Genome Project is an indication of how far biology has progressed. Barely fifty years ago, scientists were first discovering the structure of DNA. They now are in possession of an entire encyclopedia of human genetic information, and although they are not yet exactly sure what the content reveals, scarcely a week goes by without a researcher announcing a medical discovery that was made possible by the availability of the complete human genetic sequence. Coupled to this are the advances in the development of pharmaceuticals and treatments that were unheard of less than a decade ago.

But these benefits to society do not come without a cost. The terms stem cells, cloning, and gene therapy no longer belong to the realm of science fiction. They represent advances in the sciences that may hold the key to increased longevity. However, in many cases they also produce ethical and moral questions of society: Where do medical researchers obtain the embryonic stem cells for their work? Who will determine if humans can be cloned? What are the risks of transgenic organisms produced by gene therapy? These are just a few of the potential conflicts that face modern society. Only a well-educated general public can intelligently survey the pros and cons of an ethical or moral decision regarding medical science. Armed with information, concerned people can participate in the democratic process of informing their elected officials of their concerns. Science education is an important aspect of citizenship, and thus the need for series such as this to present information to the general public.

This volume covers the biology of the lymphatic system. Many people think that the sole purpose of the lymphatic system is to protect the body by way of the immune response. In reality, the lymphatic system represents a diverse system of cells, tissues, and organs, which interact with a number of other body systems. The lymphatic system serves as a second circulatory system in the body, and while it does not contain red blood cells, it does serve an important role in moving fluids. The lymphatic system is also responsible for the transport of fat-soluble nutrients from the digestive system. However, one of the more important roles is that of the immune response. Not only is the lymphatic system well equipped to protect against incoming pathogens, it also possesses the ability to destroy infected cells of our own body, thus limiting the spread of an infection. Diseases such as HIV and sudden acute respiratory syndrome (SARS) often bring attention to human disease, and initiate or escalate methods of protecting the public, such as educational classes and vaccination programs. All of us are subject to the effects of disease, and thus a knowledge of the anatomy and physiology of the human lymphatic system is important to maintain a long, healthy life.

The ten volumes of *Human Body Systems* are written by professional authors who specialize in the presentation of complex scientific ideas to the general public. Although any book on the human body must include the terminology and jargon of the profession, the authors of this series keep it to a minimum and strive to explain the concepts clearly and concisely. The series is ideal for the public libraries, as well as for secondary school and introductory college libraries. In addition, medical professionals or anyone with an interest in human biology would find this series a useful addition to their personal library.

Michael Windelspecht
Blowing Rock, North Carolina

Acknowledgments

Writing, researching, and producing this volume of the Human Body Systems Series has been both an invigorating challenge as well as a gratifying, and at times, exhausting experience. I am indebted to the editor of this series as well as my coauthor, Michael Windelspecht, for his invaluable support, assistance, and encouragement throughout the entire process. His feedback and guidance not only helped me with this project, but has helped me develop as a professional science writer. I am also deeply thankful to this volume's illustrator, Sandy Windelspecht, who was able to make art from my rough ideas and managed my repeated revisions and change suggestions. Thanks also is due to Elizabeth Kinkaid for gathering the photos and supplemental graphics. In addition, I am grateful for all the support and encouragement from Debby Adams at Greenwood, who was wonderful at dealing with my barrage of e-mails and questions.

Finally, I would like to thank my parents, Douglas and Sharon McDowell, and my sister, Christine, for their endless supplies of love, friendship, and support.

Julie McDowell

As the series editor and one of the authors in this series, I would like to thank several people without whose assistance this project would never have been realized. First, thanks to my editor, Debby Adams, for envisioning this series and keeping me focused on the big picture. An editor like Debby is invaluable to a writer and I greatly appreciate her confidence in my abilities.

As always, thanks to my wife Sandy. Not only did she serve as my support for this work, but she also held the dual role of graphic illustrator for the series. Her patience is unbelievable. In many cases I simply could not envision the final illustrations until we had undergone several revisions. I would also like to thank Liz Kincaid for her work with the illustrations, and the staff of the Westchester Book Group for putting together the final product. The production of a series such as this is a team effort, and I was working with the best.

I would also like to thank my fellow authors in the series. I was fortunate to work with some of the best in the business. While I may have been the series editor, in many cases I was learning from them.

Michael Windelspecht

Introduction

The natural world is an exceptionally hostile place, with pathogenic and parasitic organisms waiting to exploit any weakness in an organism. Fungi, bacteria, parasitic worms, protistans, and viruses abound in the natural world. The concept of survival of the fittest extends from the lowest life forms to the complex environments of primates. In order to survive all organisms must possess some mechanism of combating invaders. Humans are no exception to this rule. They are in a biological arms race with the microscopic world. Luckily, humankind possesses one of the most elaborate defensive systems on the planet—the lymphatic system.

The primary task of defending the approximately 63 trillion cells of the body against this onslaught of invaders rests with the lymphatic system. While other systems do provide some protection, such as the acids of the stomach and the structure of the skin, it is the job of the lymphatic system to initiate an immune response against invading pathogens. The lymphatic system is the system of the body that is responsible for the immune response. This tiered system of defense utilizes physical barriers, such as the skin, and general defense mechanisms, such as the white blood cells. But perhaps the most significant weapon in its arsenal is the specific defense system. In this aspect of the immune response, specialized cells called *lymphocytes* detect specific invaders (such as fungi, bacteria, and viruses) and eliminate them from the body. This response can be directed against both free pathogens in the body or against cells that have become infected. As an added protection, the specific response has the ability to "remember" an infection, practically ensuring that you will never be infected by the same organism or virus twice.

The lymphatic system does have other roles in the body. First, it acts as a second circulatory system. The lymphatic system is responsible for returning the fluid from the tissues of the body, called *interstitial fluid*, to the circulatory system. In this regard, the lymphatic system helps regulate water balance, ensuring not only that the tissues have proper fluids, but that excess fluids do not accumulate in the extremities. The second—often overlooked—role is that of a transport system. The lymphatic system moves fat-soluble nutrients from the digestive system to the circulatory system using a special class of molecules called the *lipoproteins*.

Unlike other body systems, such as the digestive system and endocrine system, the lymphatic system does not have a large number of organs dedicated to the role of immune response. While there are some, such as the thymus and spleen, the majority of the lymphatic system consists of small ducts, minor glands, and specialized cells located in other body systems. For example, in the liver a group of cells called the *hepatocytes* dominates the tissue. Located among the hepatocytes are the Kupffer cells, which act as the representatives of the lymphatic system, protecting the tissue from bacteria moving from the digestive tract.

A baby receiving a vaccination. © NMSB/Custom Medical Stock Photo.

In the study of the human body, the lymphatic system was one of the last body systems to be formally identified. While the processes of circulation and digestion had at least been documented since the time of the ancients, the lymphatic system had historically been considered to be a part of the circulatory system. Because an understanding of the immune response requires a recognition of the role of the cell in human physiology, it was not really until the nineteenth century that scientists began to recognize that there were cells whose function was to remove pathogens from the tissue. The majority of discoveries regarding the immune response were developed in the twentieth century, and some may argue that we still lack a complete understanding of how this body system works. But the progress in the past century has been staggering. A little more than two centuries ago, Western medicine began to experiment with the process of vaccinations; now vaccination programs have effectively wiped some diseases from the face of the planet. Less than a century ago, bacterial infections were often fatal, but the development of antibiotics in the twentieth century has practically eliminated the fear of death by bacterial infection. Our history as a species is linked to our success against diseases.

The study of the lymphatic system plays an important role in modern science and represents an area of intense research by private corporations, medical institutions, and academic investigators. Despite the progress, challenges still remain. Cures for the human immunodeficiency virus (HIV) still evade modern medicine, and periodic outbreaks of rare diseases, such as the sudden acute respiratory syndrome (SARS), provide a sobering reminder that, despite our wealth of medical knowledge, we still do not have a grasp on all of the possible diseases that may affect our society. Despite the success of antibiotics, we are often made aware that the war continues, because some bacteria have evolved resistance to all of the major antibiotics. The twenty-first century may be the time when we can effectively address these problems and end the threat of many diseases.

The information contained within this volume is descriptive, and not intended as a diagnostic tool for any specific disease or ailment. Medicine is a rapidly evolving field, and your physician or qualified medical professional should be consulted for any medical condition. Even something as simple as starting a new vitamin supplement should begin only after a consultation with a doctor. Only medical professionals have access to the latest news on drug interactions and treatments for diseases.

This volume is part of a ten-volume series on the human body systems. It is designed as a reference volume for anyone interested in obtaining an overview of the physiology, history, and ailments of the lymphatic system. Following a list of interesting facts about the lymphatic system, the text of this volume is divided into three parts. The first section (Chapters 1 to 5) examines the basic anatomy and physiology of the lymphatic system, from the cells of the system to the major organs and glands. Included in this is a discussion of how the lymphatic system develops before birth. The second section (Chapters 6 and 7) is a discussion of the history of discovery and study of the lymphatic system and the immune response, from the ancients to the present. In the third section (Chapters 8 to 11), ailments and diseases of the lymphatic system, including autoimmune diseases and worldwide problems such as HIV, are discussed. A chapter at the end of this section discusses the relationship between nutrition and the immune response. At the end of the volume is a list of commonly used acronyms, a glossary of important terms, a list of organizations and Web sites, and a bibliography. A general index is also provided at the end to facilitate cross-referencing of topics.

This work is targeted at the general science audience and, as such, an attempt has been made to describe as many of the medical and scientific concepts in common language. The glossary at the end of the work provides definitions or examples of key terms. **Bold** type indicates the first use of a glossary term in the text. The organization of this volume and series makes this work attractive for secondary school libraries, undergraduate higher education colleges, and universities where students may be seeking general information on the lymphatic system. In addition, community libraries that wish to possess a general reference volume on the lymphatic system, as well as anyone with an interest in science, history, or medicine, will find this work a useful addition to their collection.

INTERESTING FACTS

▶ The lymphatic system returns about 3.17 quarts (3 liters) of fluid each day from the tissues to the circulatory system.

▶ The average macrophage can engulf 100 bacteria a second.

▶ A plasma cell (B cell) can produce over 2,000 antibodies per second.

▶ The thymus gland, the site of T cell maturation, reaches its maximum size when a person is age 12, then decreases in size with age.

▶ A cubic millimeter of blood can contain up to 10,000 leukocytes, of which up to 70 percent are neutrophils and 25 percent lymphocytes.

▶ In patients with leukemia, the total white blood cell content per 0.03 ounces (1 cubic millimeter) of blood may reach over 500,000.

▶ Vaccinations against smallpox have been in use since the time of the ancient Chinese civilizations.

▶ 7,000 new cases of Hodgkin's disease are diagnosed annually.

▶ Almost 54,000 new cases of non-Hodgkin's lymphoma are diagnosed annually in the United States.

▶ Since 1981 there have been over 774,000 cases of HIV in the United States. More than 448,000 of these individuals have died.

▶ 40,000 new cases of HIV are reported in the United States annually.

▶ Some forms of elephantitis may be caused by microscopic parasitic worms that block the lymphatic vessels in the legs.

▶ Approximately 50 million Americans (almost one in six) suffer from allergies.

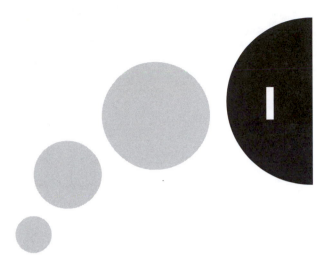

The Chemicals and Cells of the Lymphatic System

The lymphatic system is a complex group of cells, tissues, and organs that are widely dispersed throughout the human body (Figure 1.1). The lymphatic system has three primary functions. First, its cells are primarily responsible for the immune response of the body. For this reason the lymphatic system is frequently called the immune system. Most people are familiar with the immune system as it provides resistance to disease. Modern diseases such as acquired immunodeficiency syndrome (AIDS) and sudden acute respiratory syndrome (SARS) greatly challenge the capabilities of our immune system (see Chapter 10). Second, the vessels of the lymphatic system actually represent a separate circulatory system in the human body. Unlike the cardiovascular circulatory system, the lymphatic system does not directly supply nutrients or oxygen to the tissues of the body, but rather is primarily involved in the return of fluids from the tissues. Finally, the lymphatic system is involved in the transport of select nutrients from the digestive system to the circulatory system.

This chapter gives an overview of the molecules, cellular components, and chemical signals of the lymphatic system. It focuses primarily on those aspects that are associated with the immune response, although some transport molecules are also discussed. The interaction of these cells, signals, and molecules to create an immune response will be covered in the following chapters.

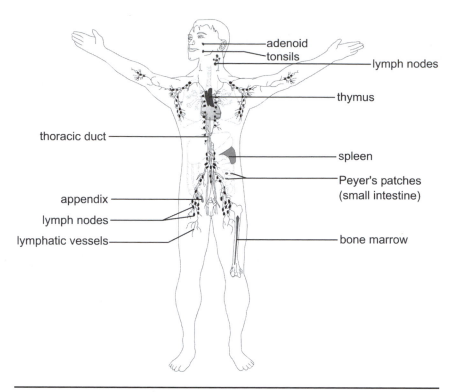

adenoid
tonsils
lymph nodes
thymus
thoracic duct
spleen
Peyer's patches
(small intestine)
appendix
lymph nodes
lymphatic vessels
bone marrow

Figure 1.1. Location of the organs of the lymphatic system.
In many cases the organs of the lymphatic system have dual roles in the body.

SUBCELLULAR COMPONENTS OF THE LYMPHATIC SYSTEM

Complement Proteins

As the name suggests, these proteins complement or assist in the function of the immune response. These are nonspecific components of the lymphatic system, meaning that they do not recognize specific types of pathogens entering the body, but instead target any form of invading bacteria or fungus. From an evolutionary perspective, complement proteins probably represent the simplest and oldest form of immune system. Forms of complement proteins are found in all animals. There are approximately twenty different types of complement proteins in humans; collectively they are called the *complement system*. Complement proteins move throughout the circulatory system in an inactive form, commonly called a *zymogen*. The mechanism by which they are activated is dependent upon the class of the complement protein. Although complement proteins are found in the circulatory system, they are considered to be part of the lymphatic system due to their association with the immune response.

The complement proteins may either target invading fungal and bacterial cells directly, or they may be recruited by antibodies or other cells of the immune system. The proteins have a variety of functions. Some classes are involved with attacking the membrane of the invading pathogen causing it to lyse, or break. Other classes interact with the antibodies secreted by the B lymphocytes (see later section, "Lymphocytes"). This is often called the *classical pathway*, because it is the most common mechanism of complement system activation. Once activated by an antibody, the complement proteins form a pore in the membrane of the invading cell, causing it to lyse. (The interaction of the complement proteins with antibodies is discussed in more detail in Chapter 4.)

Some complement proteins act as molecular flags. This class sticks to the surface of the pathogen, but rather than causing the membrane to rupture, these proteins signal macrophages and other **phagocytic cells** of the immune system to envelop the invading cell and destroy it. Other classes of complement proteins are involved in the inflammatory response (see Chapter 4) or in activating enzymes in the blood.

The complement proteins that directly lyse the membrane of the pathogen do so by what is called the *alternative pathway*. In this case, the inactive proteins are activated by some component of the bacterial or fungal **cell wall**. Once activated, the proteins congregate on the invading cell and form a pore through the membrane, disrupting the membrane barrier of the cell and causing it to lyse.

While complement proteins may appear to be an effective mechanism of immune response, they lack the ability to target specific types of cells that are invading the body. The task of targeting specific invaders falls to the cells of the immune system (see later section, "Cells of the Lymphatic System and Immune Response").

Chylomicrons and Lipoproteins

One aspect of the lymphatic system that is not involved in the immune response is the transport of fat-soluble material from the digestive tract. This includes not only **triglycerides**, but also the fat-soluble vitamins. These **hydrophobic** molecules are packaged within the small intestine into spherical structures called **lipoproteins**.

Lipoproteins are a combination of fats and proteins. Following enzymatic digestion in the lumen (cavity) of the small intestine, fatty acids are reassembled into triglycerides in the epithelial cells of the small intestine. They are then packaged into *chylomicrons*. Chylomicrons represent one form of lipoprotein that is manufactured within the lining of the small intestine. Due to their size and hydrophobic characteristics, chylomicrons can't pass into the capillaries within the villi of the small intestine, and thus are unable to be transported to the liver in the same manner as the majority of nutrients. Instead, they enter into the lacteals (see Chapter 3) of the digestive tract. The presence of fats following a meal makes the lacteals swell in size, a trait that enabled the lacteals to be first discovered as far back as the time of the ancient Greeks (see Chapter 6).

Once in the lacteals of the intestines, the chylomicrons utilize the lymphatic system to bypass the liver and travel to the heart via the thoracic duct (see Chapter 2), where they enter into the blood stream. At this point the vitamins and energy-rich nutrients within the chylomicron are removed by the tissues, and the chylomicron becomes an empty shell. The other lipoproteins, such as low-density lipoproteins (LDLs) and high-density lipoproteins (HDLs) are manufactured by liver tissue and do not enter the lymphatic system. (The role of HDLs and LDLs is covered in the Digestive System volume of this series.)

Antimicrobial Proteins

The surface cells of the body, called the *epithelia*, are most often the first to experience an attack by an invading organism. For this reason many of the body's surfaces secrete antimicrobial proteins or enzymes. An enzyme is a chemical compound (usually a protein) that accelerates a chemical reaction. Although enzymes are most often thought of in association with the digestive or nervous systems, in fact they are active in all of the systems of the body.

The surfaces of the eyes and mouth, because they are moist environments and are warmer areas of the body, represent an ideal location for a microbial attack. At these locations, the body secretes an enzyme called **lysozyme** in the saliva and tears. Lysozyme acts by degrading the cell walls of invading bacteria. Because animal cells lack cell walls, they are not disturbed by the presence of the enzyme.

This is not the only example of antimicrobial compounds in the body. Technically, the **protease** enzymes of the stomach may be considered a part of the immune response, because, in cooperation with the hydrochloric acid of the stomach, they inhibit the activity of pathogenic organisms. In the small intestine, specialized cells called *Paneth cells* secrete an antimicrobial compound called *cryptidin*. Even the bacteria located within the large intestine assist with patrolling against incoming pathogens. *Escherichia coli* (commonly called just *E. coli*), frequently considered to be a pathogen itself, helps protect the large intestine by secreting a chemical called *colicin* that prevents growth of pathogenic organisms.

These antimicrobial systems are not designed to completely prevent an attack by a pathogenic organism. Instead, like the complement proteins, the antimicrobial substances noted in this section act to slow the growth of an invader and give the specific defense mechanisms (lymphocytes) time to prepare. In this regard, antimicrobial systems are very effective in their mode of action.

LYMPHATIC FLUID

The fluid content of the lymphatic system is actually derived from the circulatory system. In the circulatory system, the capillaries represent the location where gas and nutrient exchange is most likely to occur with the surrounding tissue. Capillaries are fragile structures, whose walls are typically only one cell thick. However, these cells, called *endothelial cells*, do not form a solid structure, like that of a hose. Instead, there are small pores between the cells that form the lining of the capillaries. These pores are too small to allow the cells and plasma proteins of the circulatory system to pass, but large enough to allow a free exchange of fluid with the surrounding tissues. This fluid represents the medium through which nutrients and gases may be exchanged. The fluid, called *interstitial fluid*, bathes most tissues of the body. Cells typically deposit waste in the interstitial fluid for pickup by the circulatory system, and receive nutrients and gases to conduct their metabolic processes.

The majority of this fluid is reabsorbed back into the capillaries. However, this process is only about 85 percent effective. Each day about 3.17 quarts (3 liters) of fluid is not reabsorbed back into the capillaries, but instead remains in the tissue. This amount may not sound significant, but in an average adult there is only 5.28 quarts (5 liters) of blood. It would seem that the loss of fluid from the capillaries would represent a severe challenge for the circulatory system, and the organism as a whole. The lymphatic system makes up the difference by recycling the interstitial fluid and returning it back to the circulatory system. In most people, the lymphatic system returns around 3.17 quarts (3 liters) of fluid daily. In other words, the output of the circulatory system to the tissues is matched by the input of interstitial fluid from the lymphatic system.

Lymphatic fluid does not contain red blood cells, and in general lacks any pigmentation. However, despite its lack of color, there are plenty of ions, molecules, and cells in lymphatic fluid. These include ions such as sodium (Na^+) and potassium (K^+), chylomicrons, and a host of cells associated with the immune response (see next section).

CELLS OF THE LYMPHATIC SYSTEM AND IMMUNE RESPONSE

The immune system utilizes a number of different cell types to protect the body from infection. The major classes of cells are listed in Table 1.1. These

TABLE 1.1. Cells of the Immune System

Cell Class	Type	Specific or Nonspecific Defense	General Function
Lymphocytes	Natural killer (NK)	Nonspecific	Targets virus-infected cells
	T Cells	Specific	Attacks antigen-presenting cells
	B Cells	Specific	Produces antibodies to attack free antigens
White blood cells	Macrophages	Nonspecific	General phagocytic cells
	Neutrophils	Nonspecific	One-time-use cells that contain powerful chemical reactions
	Eosinophils	Nonspecific	Destroys parasitic organisms
	Basophils	Nonspecific	Releases histamine
	Mast cells	Nonspecific	Releases histamine

cell types may either be generalists (nonspecific defense mechanisms) or specialize in the destruction of certain identified invaders of the body. Cells of this second class belong to the specific defense mechanism of the body.

The lymphatic cells are derived from the same type of cell in the bone marrow as the cells of the circulatory system. The common name for this type of cell is called a *stem cell*. While stem cells are commonly thought of as being able to form any type of cell in the body, in reality they vary in this ability. Some stem cells, such as those in early embryonic development, are totipotent, meaning that they have the ability to form virtually any cell type. However, shortly after the embryo starts to develop, stem cells lose their potency, and thus their ability to form certain types of cells. The stem cells in the bone marrow are pluripotent cells, indicating that they are limited in what types of cells they can differentiate into. The type of stem cell that gives rise to the immune cells, as well as the majority of cells in the circulatory system, is called a *hematopoietic stem cell*. From this stem cell are derived progenitor cells, which possess an additional layer of specialization. Two different types of progenitor cells are involved with the formation of lymphatic cells. The lymphocytes are derived from the lymphoid progenitor cell, while the leukocytes and macrophages are derived from the myeloid progenitor cell. The cell types also differ in where they mature in the body and their contributions to the function of the lymphatic system and immune response.

Natural Killer Cells

Natural killer (NK) cells are another example of a nonspecific defense mechanism in the body. While the complement system acts nonspecifically against invading fungal and bacterial cells, the role of the NK cells is to eliminate cells of the body that have either been invaded by **viruses** or are cancer cells. It is important to note the difference between NK cells and a form of T cells called the *cytotoxic T cells* (see next section). While both attack viral infected cells and cancer cells, cytotoxic T cells are specific in their targets, meaning that they will only destroy cells that have been infected with a specific virus. NK cells are generalists and will destroy any viral infected cell that they come in contact with. Natural killer cells are not phagocytic cells, but rather destroy the target cell by lysing the membrane. Natural killer cells belong to a class of cells called the *lymphocytes*, of which the T and B cells are the most commonly recognized. NK cells are formed in the same manner as T and B cells (see next section), but do not mature in the **thymus**, as is the case with the T cell.

Lymphocytes

The term *lymphocyte* is most commonly used to describe two groups of lymphatic cells, the B cells and T cells, although, as noted in the preceding section, the NK cells also belong to this class. Lymphocytes start as hematopoietic stem cells in the bone marrow of the long bones of the body. The hematopoietic stem cells form progenitor lymphoid cells, which then divide into the cell lines that will form the B cells, T cells, and NK cells. Unlike B and T cells, NK cells do not require additional processing and instead proceed directly into action as nonspecific defense mechanisms.

B and T lymphocytes are named for the location in the body in which they complete their maturation process. An immature T cell migrates to the thymus to finish its development; a B cell completes its maturation in the bone marrow. The "B" does not actually stand for "bone," but rather a structure called the *bursa of Fabricus*. This structure is only found in birds, but is where the B cells were first discovered (see Chapters 6 and 7). Since the B cells of all other vertebrates mature in the bone marrow, the B is commonly considered to refer to "bone." Although both B cells and T cells are lymphocytes and are involved in the defense of the body against specific pathogens, their modes of action are very different.

T and B cells both respond to specific antigens in the human body. An **antigen** is a molecule that invokes an immune response. All cells and viruses have unique antigens present on their surface. What distinguishes the cells of our body from invading viruses and the cells of invading bacteria, **fungi, protistans**, and parasitic worms is the presence of self-markers (see later section, "Identification of Self: The Role of Cellular Markers"). In other words, if a cell cannot identify itself as a normal part of the human body, it runs the risk of initiating an immune response.

The role of the B cells is to develop antibodies against antigens that present themselves in the tissues and fluids of the body. Antibodies are proteins that target the antigen and either mark it for destruction by nonspecific mechanisms or physically destroy the molecule. (The action of antibodies will be covered in greater detail in Chapter 4). Because almost anything may be an antigen (proteins, cellular debris, chemicals, etc.), it is possible for a B cell to mistakenly identify a cell of the body as an antigen. Therefore, immature B cells are screened while in the bone marrow before maturing and being sent to secondary lymphoid tissues (see Chapter 3), such as the appendix and lymph nodes. This process is often called *self-tolerance*, because the B cell must be able to tolerate the wide range of potentially false antigens that are produced by the cells of the body. However, sometimes this screening is not completely effective, resulting in an autoimmune disease. (The different types of autoimmune diseases are covered in Chapter 10.)

Each B cell of the body will recognize—and produce antibodies against—one specific antigen. Antibodies are proteins, and are manufactured in the same manner as other proteins in the body. The instructions for producing the antibody are stored as **genes** in the deoxyribonucleic acid (DNA). When needed, these genetic instructions are transcribed into a message (called messenger ribonucleic acid, or mRNA), which proceeds to the cytoplasm of the cell to be translated into a functional protein (see Figure 1.2).

All antibodies have a characteristic structure as shown in Figure 1.3. Each antibody contains two light chains and two heavy chains. They are called the light and heavy chains based upon the number of amino acids in the peptides that make up their structure. The chains are held together by disulfide bonds, forming a "Y" shaped molecule. There is very little variation in the constant regions (or C regions) of the heavy chains in the antibody. In humans, there are just five major variations in this area of the heavy chain, which correspond to the five major classes of antibodies (see Chapter 4). It is the combination of light and heavy chains that is responsible for the tremendous variation in antibody specificity. At the terminal end of each chain is an area called the *hypervariable segment*, which is ultimately responsible for targeting a specific antigen. The mechanism by which these hypervariable regions in the chains are generated is one of the amazing features of the immune system.

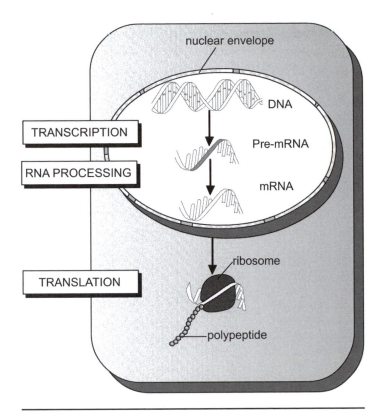

nuclear envelope

DNA

TRANSCRIPTION

Pre-mRNA

RNA PROCESSING

mRNA

ribosome

TRANSLATION

polypeptide

Figure 1.2. The central dogma of biology.
The process by which the genetic information is read to manufacture a functional protein.

The structure of the antibody, and thus its effectiveness in recognizing the correct antigen, is determined by the sequence of genetic information that is used to construct the protein. Even though the human genome contains over 3 billion pieces of information, organized into over 30,000 genes, there is still not enough information or room to have a single gene for each antibody that a human being may require over an entire lifetime. It is believed that the immune system has the potential to produce between one million and one billion different types of antibodies. It is simply not possible that each antibody is the result of a single dedicated gene in the DNA.

It is now known that there are only a few hundred genes that are responsible for generating the diverse array of antibodies. These genes are grouped into four major types (see Figure 1.4):

- Genes grouped into the C regions are responsible for generating the protein sequences in the constant regions of the heavy and light chains.

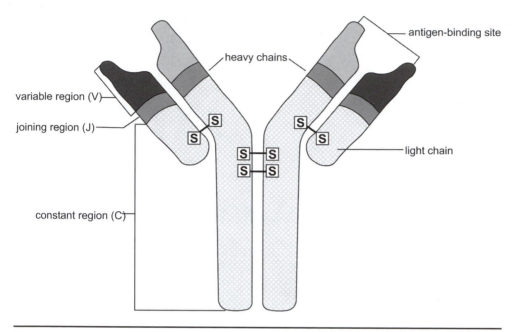

Figure 1.3. Antibody structure.
This diagram illustrates the basic structure of an antibody protein.

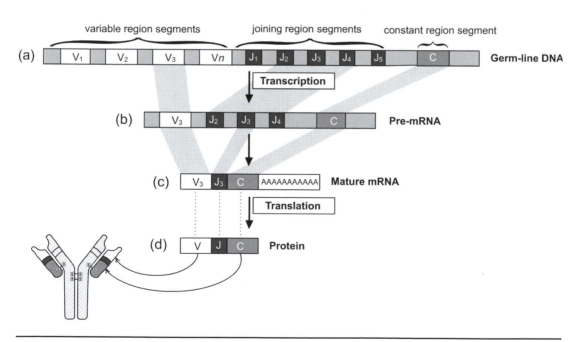

Figure 1.4. The genetic basis of antibody diversity.
This diagram simplifies the genetic mechanism responsible for the production of antibodies by a B cell. The choice of a specific V or J region results in a minor change in the structure of the final antibody.

- The J genes are responsible for generating the small peptide segments that link portions of the antibody together.

- The D group of genes encodes for a small diversity region found only in the heavy chain.

- The V genes provide the information for generating the hypervariable segments of the light chains.

It is important to note that the following process of selecting the genes that will form an antibody occurs *before* the B cell encounters an antigen. In other words, as the B cell matures, it becomes specific in what type of antigen it will respond to. It is extremely possible that a B cell may never come in contact with its antigen, and thus never be involved in an immune response. However, the large number of B cells in circulation, and the mechanism of the specific immune response, means that the body needs only one B cell that recognizes the antigen to mount an effective immune response. Because many invading pathogens, such as bacteria and viruses, may present multiple antigens, it is possible that more than one B cell may be producing antibodies for the same pathogen at the same time (this will be covered in more detail in Chapter 4).

To construct a light chain (Figure 1.4), it is necessary to have a single V, J, and C gene segment (there are no D segments in the light chain). There are 30 to 40 V segments, 4 to 5 J segments, and a single C segment in the area of DNA responsible (located on human chromosomes 2 and 22) for the formation of the light chain. This alone produces around 200 different combinations. Furthermore, the regions are subject not only to mutation, but also to minor variations in the reading and processing of the genetic information. These mistakes serve to increase the variation in the segments. Thus it is highly unlikely that two B cells will be identical in the types of antibodies that they produce. Before maturation, the lymphocyte possesses all of the gene segments. By a process called *somatic recombination*, a single V and J region are selected and then matched up to the one C region. After translation (Figure 1.3) the result is a protein (also called a *peptide*) that will become the light chain of the antibody.

The synthesis of a heavy chain is slightly different, but follows the same general pattern. There are 65 V segments, 27 D segments, and 5 J segments available on chromosome 14 to construct a single heavy chain, resulting in 10,530 possible combinations. Once again, a single V, D, and J segment are combined, and then linked to a C segment. The same errors that produced variation in the light chains may also play a role in the formation of the heavy chains, resulting once again in an almost endless source of variation in heavy chain structure. Following processing, transcription of the information, and translation into a functional protein, the result is the heavy chain of the antibody. The light chains and heavy chains are then linked together, forming an antibody. (The roles of the different antibody types in the immune response will be covered again in Chapter 4 as part of the humoral immune response.)

Although all B cells undergo a similar maturation process, there are minor variations in their form and function in the immune system. Once activated by a specific antigen (see Chapter 4), a B cell rapidly divides, forming a large number of effector B cells or plasma cells. It is these cells that actively combat the antigen in the body in what is called the *primary immune response*. As the primary response progresses, some of the activated B cells are retired, forming memory B cells. It is these cells that are responsible for the secondary immune response that occurs when the body is exposed to the same antigen later in time. All B cells are involved in the humoral response (see Chapter 4), which targets free antigens in the system.

As noted, T cells complete their maturation in the thymus, one of the primary lymphoid organs of the body. The thymus is located just above the heart. (The structure of the thymus is covered more extensively in Chapter 2 and its embryonic development is discussed in Chapter 5.) In the thymus, the immature T cells undergo a series of modifications. The most important of these changes occurs as specific genes within the T cells are activated, which enables the production of unique proteins, called **glycoproteins**, on the surface of these cells. These proteins, examples of which are called CD4 and CD8, play an important role in the function of the immune system (see later section, "Identification of Self: The Role of Cellular Markers"). As was the case with the B cells, the maturing T cells in the thymus are screened to ensure that they are not recognizing any of the tissues or cells of the body as invading antigens. Those T cells that display an affinity for self are targeted for cell death and usually do not mature.

While T cells are specialized for the targeting of antigen-presenting cells of the body, there are actually several different forms of T cells, each of which has a specific function in the immune response:

- *Helper T cells.* Helper T cells serve as the liaison between the nonspecific and specific defense mechanisms. They are responsible for activating both the humoral and cell-mediated responses in the body, by interacting with both mature T and B lymphocytes.

- *Cytotoxic T cells.* These cells destroy infected body cells under the direction of the helper T cells.

- *Inducer T cells.* Cells located in the thymus that are responsible for T cell maturation.

- *Suppressor T cells.* The task of these cells is to shut down the immune response once the infection is complete. These cells divide much slower than other T cells, producing a natural delay mechanism.

- *Memory T cells.* As was the case with the B cells, some T cells are held in reserve for future exposures to the antigen. These are called *memory cells*, because they originated with the initial exposure to the antigen.

The activity of T cells is highly dependent on cell-to-cell signaling using proteins embedded in the plasma membrane. T cells not only serve to identify cells of the body, but also to distinguish between infected and healthy cells so as to prevent unnecessary tissue damage. This identification is made possible by specific receptors on the surface of the T cell. The antigen identification portion of these receptors is highly variable. T cells use a similar method of generating variation as do the B cells with antibody formation. Within the genome of the T cell are the same variable (V), diversity (D), joining (J), and constant (C) families of genes that are present in the B cells. These genes are rearranged in much the same way as in the B cells to produce a receptor that is specific for one type of antigen recognition. As is the case with the B cell, each T cell is specific in what it can recognize. The difference is that while the antibodies of the B cell recognize free antigens, such as might be found in the fluids of the body, the receptors of the T cells are designed to identify cells that are presenting a specific antigen as the result of being infected by a pathogen, such as a virus. This is a complex interaction, which involves a number of cellular markers and proteins. (The role of these receptors and proteins is covered in the next section, "White Blood Cells," and again in Chapter 4.)

White Blood Cells

The term white blood cell (WBC) is used as a general description for a wide variety of cells in the immune system, including macrophages, **neutrophils**, eosinophils, and basophils (the general function of each is provided in Table 1.1). A white blood cell begins as a hematopoietic stem cell in the bone marrow. This cell then begins to differentiate into a myeloid progenitor cell, which then becomes basophils, eosinophils, neutrophils, and a precursor of macrophages called **monocytes**. Typically, the term *white blood cell* indicates the macrophages.

Macrophages are the general workhorses of the immune system. These are amoeba-like cells that move throughout the body. Unlike many cells of the circulatory and lymphatic system, the macrophages are not confined to capillaries. Instead, they are able to move freely between the circulatory system and the interstitial fluids that bathe the tissues of the body. *Macrophage* is the collective name for these cells, but they are sometimes called by other names throughout the body (see Table 1.2). Inactive macrophages are monocytes. These may be activated at the site of an infection by a variety of mechanisms.

Macrophages are phagocytic cells, meaning that in order to destroy pathogens, macrophages must first ingest them. Phagocytosis involves a budding in of the cell membrane, forming a vesicle inside the cell. Once the pathogen is engulfed, the macrophage can utilize two major mechanisms for destroying the invader. First, the macrophage may merge the pathogen-containing vesicle with another cellular vesicle called a *lysosome*. The lyso-

TABLE 1.2. Nomenclature for Macrophages in the Body

Tissue	Name
Digestive system (liver)	Kupffer's cells
Urinary system (kidney)	Mesangial cells
Connective tissue	Histocytes
Respiratory system (lungs)	Alveolar macrophages
Nervous system (brain)	Microglial cells

some contains powerful digestive enzymes, which effectively "eat" the pathogen, rendering it harmless. A second mechanism involves making the internal environment of the vesicle containing the pathogen extremely toxic. Macrophages can produce free radicals such as nitric oxide and superoxide anion. These chemicals are highly destructive to organic molecules, and quickly destroy the incoming pathogen.

Once the pathogen has been destroyed, an interesting change happens in the receptors of the macrophage. The macrophages are nonspecific generalists, but they play an important role in informing the lymphocytes (specific defense) of the presence of a pathogen in the body. Small pieces of the pathogen are moved to receptors on the cell membrane. These receptors, part of the self-identification process (see next section, "Identification of Self: The Role of Cellular Markers"), act as an activation mechanism for the helper T cells, an important link between the specific and nonspecific responses. Once the macrophage alters the surface receptors, it is then called an *antigen-presenting cell* (APC; see Chapter 4).

In many ways, the neutrophils are very similar to the macrophages. Both are phagocytic cells that patrol the body looking for pathogenic organisms or viruses. Neutrophils are actually the most abundant type of white blood cell in the body. They are typically found only at the site of an infection, usually because they are attracted by macrophage activity and chemicals released from damaged cells. Unlike the macrophages, which rely almost exclusively on phagocytic activity, the neutrophils have a variety of options available for the destruction of incoming pathogens.

Neutrophils may engulf pathogens in a manner similar to the macrophages. However, they also possess a more destructive mechanism. Each neutrophil contains a limited number of internal structures, called *granules*. These granules contain a variety of substances that are highly toxic to microorganisms. These include oxygen radicals, antimicrobial proteins, peroxynitrate (a nitric oxide compound), and hydrogen peroxide. When a neutrophil reaches the site of an injury or infection it releases these chemicals into the surrounding environment, killing not only microorganisms but often cells of the body as well. Each neutrophil only contains a limited arsenal of chemicals, and once depleted the neutrophil dies. Often, the neutrophil may be destroyed by the very chemicals that it releases. The debris from these dead cells is what forms the pus at a site of a wound or infection.

Another form of white blood cell is the eosinophil. The eosinophils are another nonspecific mechanism, but one that is targeted against parasitic organisms, such as intestinal worms, flukes, microscopic nematodes, and even ticks and mites. Eosinophils act by orientating themselves along the surface of the parasitic invader and then releasing chemicals to destroy the membrane or surface of the organism.

The last two major types of white blood cells are not directly involved in the destruction of invading organisms, but rather are involved in some of the chemical signaling. These are the basophils and mast cells, which are responsible for the release of a chemical called *histamine*, an important chemical in the inflammatory response (see Chapter 4). The role of these cells in histamine production is discussed in the last section of this chapter, "Chemical Signals in the Lymphatic System."

IDENTIFICATION OF SELF: THE ROLE OF CELLULAR MARKERS

Before beginning a discussion of cellular markers, it is first important to understand the basic structure of a cell membrane (see Figure 1.5). The membrane of a cell is composed primarily of a molecule called the *phospholipid*. Phospholipids belong to a class of **biomolecules** called the lipids and are unique in that they contain both **hydrophilic** and hydrophobic regions. When these molecules are placed in an aqueous environment, the hydrophilic and hydrophobic regions align to form a double layer, called a *lipid bilayer*. This lipid bilayer forms the basic structure of the cell membrane, the properties of which effectively block the passage of most molecules into the cell.

Located within the phospholipid layer of the cell membrane are a wide variety of proteins. Some of these proteins serve as channels through the membrane, while others act as receptors for chemical signals passing back and forth between the tissues of the body. The proteins of interest in the immune response belong to the glycoproteins, the proteins that have a sugar group attached to their outer surface. Glycoproteins are common on the surface of the cell membrane, but two types of glycoproteins play an important role in the immune response. These are the major histocompatibility complexes (MHC) markers.

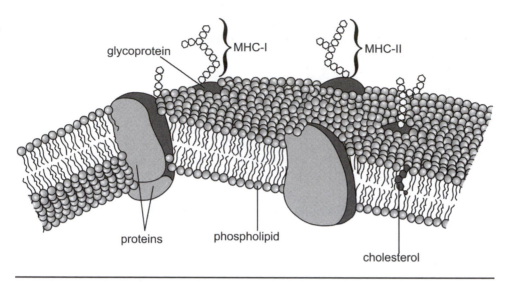

Figure 1.5. The plasma membrane of a cell.
This illustration focuses on the membrane proteins that play a major role in the immune response.

MHC markers should be considered a form of cellular identification card. All of the cells of the body have the same identification tag, which enables the body to distinguish between "self" and invading microorganisms and viruses. There are two classes of MHC markers in the cell, called MHC-I and MHC-II. MHC-I markers are more general, and are found on every cell of the body. MHC-II markers are slightly more specific, and are found almost exclusively on cells of the immune system.

There is a tremendous amount of variation between individuals in the structure of the MHC markers. While there are only three genes responsible for forming an MHC-I protein (all on human chromosome 6), there exists a number of variations, or **alleles**, for each of these genes. This allele effectively ensures that the MHC signature of one individual is unique.

CHEMICAL SIGNALS IN THE LYMPHATIC SYSTEM

Because the cells of the immune system are not all localized in a single tissue, the system possesses an elaborate series of chemical signals to operate effectively. These signals perform a variety of functions, from the recruitment of nonspecific defense mechanisms, to the activation of cells involved in the identification and destruction of specific pathogens. The cytokines are a special group of chemical signals in the immune system. These protein, or peptide, signals are secreted primarily by the T cells of the body to influence the activity of both the specific and nonspecific cells of the immune system. There are currently thirteen different types of cytokines that

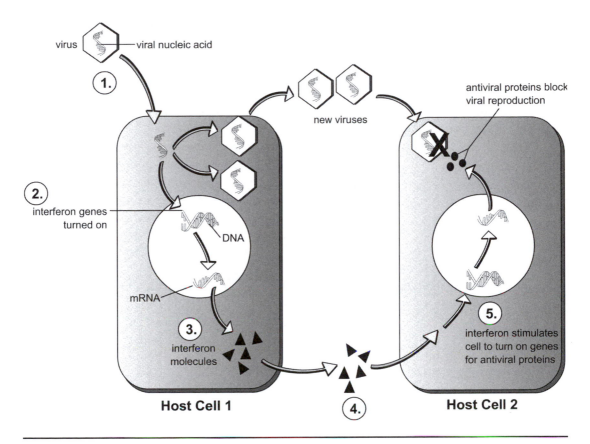

Figure 1.6. Interferon activity.
The cell on the left has been infected by a virus. It then produces interferon, which is detected by a nearby cell (on the right), enabling the production of antiviral proteins.

have been identified. The majority of these belong to two major groups, the interferons and the interleukins.

As their name implies, the interferons are involved in an interference response in the immune system. There are several different types of interferons active in the immune response. Type 1 interferons (also called *interferon alpha* and *interferon beta*) are used as a local defense against invading viruses (see Figure 1.6). When a cell is infected with a virus, it secretes interferons into the surrounding interstitial fluids. On neighboring cells, the interferon interacts with a receptor common on the surface of all cells. This causes the cell to activate antiviral protection mechanisms (frequently protease enzymes), which inhibit viral replication in uninfected cells. The type 1 interferons also enhance the development of APCs, and serve to activate NK cells in the area. It is important to note that the secretion of interferons does not protect the infected cell, only those in close proximity to it. It is also nonspecific, meaning that neighboring cells are temporarily protected against any viruses in the area.

Another type of interferon is called interferon gamma. This interferon is secreted by selected T cells and is not directly related to the type 1 interferons. The role of this interferon is to activate macrophages near the cytotoxic T cells, thus providing a better coverage in the area of an infection or wound.

Interleukins are the second major form of cytokine. There are a number of interleukins in the immune system, the most common of which are interleukin-1 (IL-1) and interleukin-2 (IL-2). IL-1 acts as a link between the nonspecific and specific defense systems. After a macrophage has engulfed a pathogen and become an APC, it secretes IL-1 to help activate helper T cells in the area. The helper T cells then communicate directly with cytotoxic T cells and B cells to begin the humoral and cell-mediated responses to the antigen.

The second major form of interleukin, IL-2, is basically an activated signal. Secreted by helper T cells, the signal activates B cells to begin antibody production, as well as cytotoxic T cells to begin destruction of infected cells of the body. The loss of helper T cells, as is the case with AIDS, means that this signal is not present, and the specific defense systems are not activated (see Chapter 4).

There are other chemical signals in the immune system besides the interferons and interleukins. In response to injury, basophils and mast cells release a chemical called *histamine*. Histamine causes the cells of the capillary beds (circulatory system) to dilate, increasing the amount of fluid (but not the amount of red blood cells) flowing out of them. This increase in interstitial fluid increases the pressure, slowing the spread of bacteria and other pathogens into the wound. In addition, clotting proteins can now move more easily to the site of the wound, allowing for a more rapid healing process. Macrophages and NK cells also benefit by the ease of moving into the interstitial spaces, allowing for a more rapid cleanup to begin. Antihistamine medications, such as those used for the common cold, reverse this process and may actually slow the immune response.

2

The Lymph System: An Examination of the Lymph Nodes and Lymphatic Circulation in the Body

One of the primary functions of the lymphatic system is to capture and collect the protein-rich fluid that escapes from the circulatory system's blood vessels and deposit it back into the tissue network. These proteins cannot be reabsorbed, therefore the lymphatic circulatory system must fetch them and bring them back. As explained in Chapter 1, the lymphatic system's circulatory functions are separate from the cardiovascular system in the human body. However, unlike the cardiovascular system, the lymphatic system's circulatory processes are focused on returning fluids from other areas of the body, rather than directly supplying the tissues with the nutrients and oxygen that they need to function.

This chapter will begin to look at some of the basic components of the lymphatic system—the lymph nodes and the elements involved in lymphatic circulation, in addition to the system's primary organs, the **bone marrow** and the thymus. These aspects of the lymphatic system are important to understand before learning how the more complex functions, such as immunity and autoimmune response, work to protect the body. See Figure 2.1 for a drawing of all the major lymphatic organs, some of which will be covered in the next chapter.

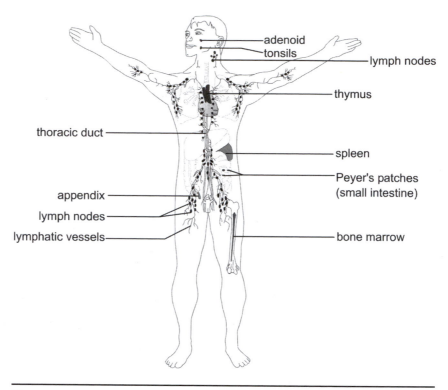

Figure 2.1. The major organs of the lymphatic system.

LYMPHATIC VESSELS

Figure 2.2 shows the relationship between the cardiovascular circulatory system and the principle parts of the lymphatic system: the lymph nodes, lymphatics, and lymph capillaries. The lymphatic vessels begin as blind-end tubes—called *lymph capillaries*—that form in the spaces between cells. Lymph capillaries are slightly larger, in addition to being more permeable, than the blood capillaries in the circulatory system. These capillaries can form in most regions of the body, and converge to form larger lymph vessels called *lymphatics*. As evident in Figure 2.2, these lymphatic vessels have a vein-like appearance, although their walls are thinner and they contain more valves than blood veins. In addition, at various spots in their structure, lymphatics contain lymph nodes (which will be discussed in detail in the next section).

Lymph is the name of the fluid that enters the lymph capillaries. As explained in Chapter 1, tissue fluid comes from the filtration in the capillaries. While the process of **osmosis** allows much of this fluid to return to the blood, some of the fluid is lodged in interstitial spaces. The lymphatic vessels return this interstitial fluid to the blood to become **plasma** again. With-

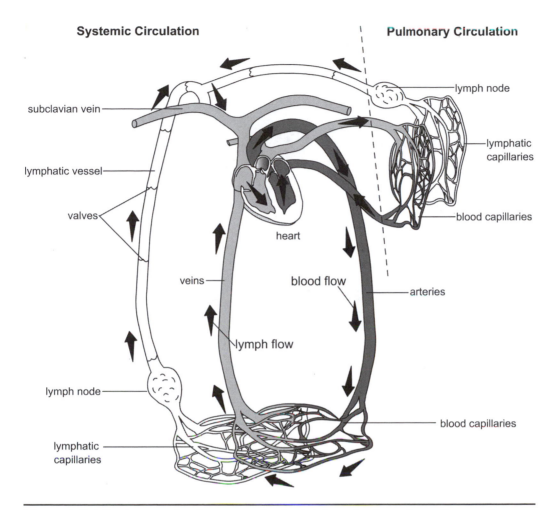

Systemic Circulation   **Pulmonary Circulation**

Figure 2.2. The interaction between the body's lymphatic and cardiovascular systems.
Tissue fluid is collected by lymph capillaries, and is then returned to the blood. The arrows indicate the flow of the lymph and blood.

out this occurring, blood volume and blood pressure would rapidly decrease, eventually leading to serious health threats, such as a heart attack or stroke.

Unlike the circulatory system, there is no pump for the lymph. In the circulatory system, the heart serves as the pump to keep blood moving throughout the body. In the lymphatic system, the lymph is kept mobile through the muscles of the lymph vessels. As the smooth muscle layer of the larger lymph vessel constricts, the one-way valvular structure prevents the backflow of the lymph.

As the lymph capillaries form lymphatics, the lymphatics eventually

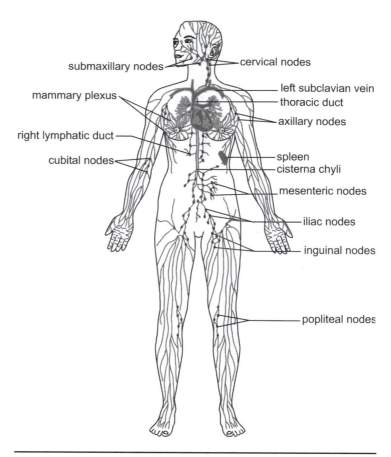

Figure 2.3. The primary groups of lymph nodes and the lymph vessel system.
The right and left subclavian veins return lymph to the body's blood supply.

merge into two main structures or channels, called the *thoracic duct* (also called the *left lymphatic duct*) and the *right lymphatic duct* (see Figure 2.3). Lymphatic vessels from the left side of the head, neck, and chest, in addition to the left upper extremity and the entire body below the ribs, all converge in front of the lumbar section of the vertebrae to form the cistern chyli vessel, which then continues to creep up the backbone as the thoracic duct. This duct then empties the fluid into the left subclavian vein, where a pair of valves are located to prevent the passage of blood into the thoracic duct. The second channel, the right lymphatic duct, takes the lymph from the right side of the body and then deposits it into the opposite, or left subclavian vein.

When doctors need to take a detailed look at the lymphatic vessels and organs, they rely on a procedure known as a lymphangiography. The lymphatic vessels and organs are filled with an opaque substance and then filmed, which produces an lymphangiogram. This image is useful for identifying **edemas, carcinomas**, and viewing any irregularities of the lymph nodes.

LYMPH NODES

The lymphatics contain structures that are oval in shape called *lymph nodes*. These bean-like organs can range in size from 0.04 to 1 inch (1 to 25 millimeters). Blood flows into lymph nodes on the way to subclavian veins. Each lymph node: contains a hilum, which is a slight depression on one side where the blood vessels enter and leave the node. Three structural elements form the framework of a lymph node: the capsule, trabeculae, and the hilum. The capsule is made up of fibrous connective tissue that not only covers the node, but also extends into it. These extensions into the node are call *trabeculae*. Inside the node, the outer cortex is composed of tightly packed lymphocytes organized into lymph nodules. These nodules contain germinal centers, where lymphocytes are actually produced. Then the inner portion of the lymph node is called the *medulla*, which contains lymphocytes that are organized into strands called **medullary cords**.

Lymph nodes contain two kinds of vessels: afferent and efferent. Lymph leaves the node through one or two **efferent vessels**, while it enters through one or a couple **afferent vessels** (see Figure 2.4). Once the lymph is in the node, any bacteria and other foreign materials it is carrying are phagocytized (consumed) by macrophages. The lymphocytes contain plasma cells that produce antibodies to counteract any pathogens that the lymph brings with it. These antibodies, in addition to the lymphocytes, will eventually travel to the blood (antibodies will be discussed in more detail in Chapter 4).

Once lymph enters the node through the afferent vessels, which are located at various places on the surface of the node, the fluid then enters the node's sinuses, which are a series of irregular channels. After passing through the afferent vessels, lymph enters the cortical sinuses and then circulates through the medullary sinuses located between the medullary cords (as noted in Figure 2.4). After the lymph passes through these sinuses, it then travels to the efferent vessels, which are located at the node's hilum structure. While the afferent vessels only open toward the node, the efferent vessels only open outward away from the node, pushing lymph out from the structure. In addition, there are fewer efferent vessels (although the actual vessels are wider) than there are afferent vessels.

Macrophages that contain phagocytic cells are located along the sinuses. Lymph travels through the nodes, and is then processed by these phagocytic

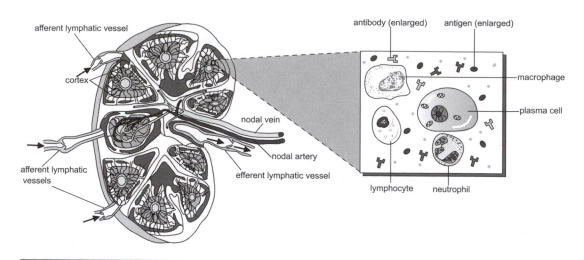

Figure 2.4. Lymph node.
On the left, a cross-section view of a lymph node, with the arrows indicating the lymph flow. On the right, a detailed view of antigen being destroyed within the lymph node.

cells, which work to separate out the bacteria, dirt, and other contaminants from the fluid. In addition, the nodes are where the lymphocytes and plasma cells are produced, which then lead to the formation of antibodies. When too many pathogens and microbes enter the node, it can become infected, causing the node to enlarge and become inflamed.

There are numerous groups, or chains, of lymph nodes located along the body's lymph vessel network, but the three primary groups of nodes classified according to their location in the body—cervical, axillary, and inguinal—are detailed in Table 2.1 and identified in Figure 2.3. Each of these node groups are located at an important junction of the body: the cervical nodes are located near the neck and head junction; the axillary nodes are located near where the arm meets the trunk of the body; and the inguinal nodes are found near where the leg meets the trunk. This is important because the skin is more likely to break in these areas, thus allowing pathogens to enter the body. For example, it is more likely that skin will break in the head, arms, or legs, rather than the trunk of the body. Therefore, if pathogens enter the body through any of these locations, the lymph will destroy them before they reach the trunk, and also before the lymph is returned to the blood contained in the subclavian veins. As these lymph nodes are processing the pathogens and bacteria, inflammation and temporary infection can occur. For instance, swollen glands frequently accompany "strep throat," which is an infection caused by a bacteria called *Streptococcus*. The glands that are swollen are actually the cervical nodes, which have temporarily enlarged as the macrophages fight off the bacteria in the throat's lymph.

TABLE 2.1. Groupings or Chains of Nodes

Group	Location and Function
Deep cervical lymph nodes	Located along the internal jugular veins, these nodes process lymph from the head and neck.
Axillary lymph nodes	Located in the chest and underarm areas of the body, these nodes process lymph from the skin and chest muscles, which include the breasts.
Inguinal lymph nodes	Located in the groin region, these nodes drain lymph from the lower extremities of the body, including the genitals.

There is also a specific kind of lymphatic tissue found in all mucous membranes, which line those systems of the body that have exterior openings to the environment. These include the respiratory, digestive, urinary, and reproductive tracts. All of these systems are lined with mucous membranes for protection. Located under the epithelial layer of these membranes are small groupings of lymphatic tissue know as lymph nodules. This is an important area for lymph nodules because, while these systems are shielded from some contamination through the mucous membranes, they are still vulnerable to coming under attack by microbes, bacteria, and various pathogens. If bacteria is inhaled and enters the body through the respiratory system, the lymph nodules located in the trachea will counteract that bacteria before it even reaches the blood. Two kinds of lymph nodules are Peyer's patches (located in the small intestine) and **tonsils** (located in the pharynx); these and other secondary lymphatic organs will be discussed in Chapter 3.

BONE MARROW

The three kinds of blood cells—white blood cells (WBCs), red blood cells, (RBCs), and platelets—are produced in two kinds of hemopoietic tissues: red bone marrow (or simply bone marrow), and lymphatic tissue that is found in the spleen, thymus gland (see the later section, "Thymus"), and lymph nodes. The red bone marrow is spongy tissue found in flat and irregular bones. Basically, the purpose of the RBCs is to carry oxygen throughout the body. Through a protein that they carry called hemoglobin, RBCs are able to bond to oxygen molecules (the function of the RBCs are covered extensively in the Circulatory System volume of this series, and WBCs have been extensively covered in Chapter 1 of this volume).

Before RBCs are produced in the bone marrow, they are stem cells that are constantly changing to form all kinds of blood cells. The rate of RBC production is high; approximately a few million are produced every second. This production rate, however, is regulated by the presence of oxygen. If plenty of oxygen is available in the body, then the bone marrow will produce the RBCs at a normal rate. However, if the body is low on oxygen (or in a state of **hypoxia**), then the kidneys will begin producing a hormone called **erythropoietin**, which causes the bone marrow to produce more RBCs, which are then able to carry oxygen through the body. Once the oxygen begins making its way through the circulatory system, the body is no longer in a hypoxic state. A person can become hypoxic following a hemorrhage or other injury that has caused them to lose a lot of blood, or if they spend a significant amount of time in higher altitudes.

As stem cells develop into RBCs, they go through a number of stages (see Figure 2.5). The most important stages are the last two: when the cell is a **normoblast** and then a **reticulocyte**. The last stage where the cell actually has a nucleus is the normoblast stage. After this stage, the nucleus disintegrates and the cell becomes a reticulocyte. While a small number of reticulocytes in the blood's circulation is normal, too many (in addition to the presence of normoblasts) could indicate that there are not enough mature RBCs available to transport oxygen. Once again, this could be due to an injury, such as a hemorrhage, or a disease.

In order for these stem cells to mature into RBCs, they need a significant amount of nutrients, such as protein and iron. In fact, protein and iron are necessary in order for hemoglobin to synthesize. In addition, vitamins such as folic acid and B_{12} are needed in order for the stem cells' genetic material to synthesize in the bone marrow. There are two necessary chemical agents called *factors* that must be present in order for the stem cells to mature into RBCs: the **extrinsic factor** and the **intrinsic factor**. The source for the extrinsic factor, also known as vitamin B_{12}, is, as the name implies, external—food. The intrinsic factor comes from certain cells, called *parietal cells*, of the stomach lining. This factor then combines with food's vitamin B_{12} resources in order to prevent the vitamin's ingestion in the stomach so it can instead be absorbed in the small intestine. If the body is not getting a sufficient supply of the intrinsic or extrinsic factors, then the person might develop **anemia**. (The role of nutrition in enhancing the immune response is examined again in Chapter 11.)

Once the RBCs are produced, they live for approximately 120 days; past this time, they lose their durability and become fragile. At that point, they are removed from the circulatory system by the tissue macrophage system. Macrophages—which are contained not only in the bone marrow, but also the liver and the spleen—are the lymphatic system's consumers. These old and failing RBCs are eaten (phagocytized) and digested by the macrophages. However, the iron in the RBCs is extracted and placed in the blood, even-

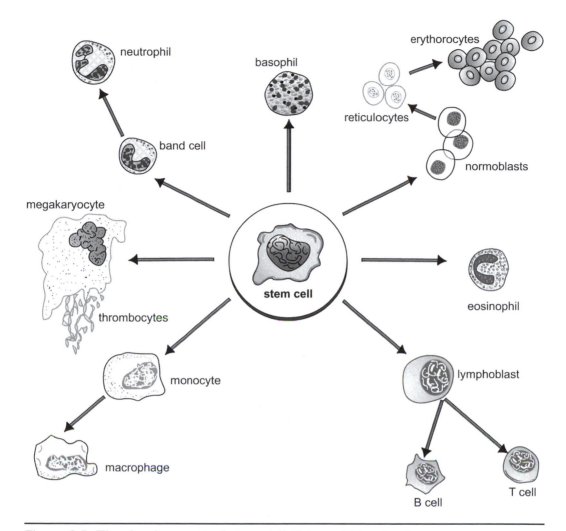

Figure 2.5. The development of stem cells into red blood cells.
Stem cells are the initial form of all types of red blood cells, and are found in both the red bone marrow and lymphatic tissue. Thrombocytes are also called platelets.

tually returning back to the bone marrow to be synthesized into new hemoglobin. This recycling process is done repeatedly. All excessive amounts of iron are stored in the liver until needed by the bone marrow.

But not every aspect of the RBCs can be recycled. As the macrophages are processing the RBCs, they take the "heme" portion of the hemoglobin and convert it into bilirubin. The liver then takes the bilirubin and excretes it into bile. The bile leaves the liver, travels to the small intestine and colon, and eventually leaves the body as the waste product known as feces. A small amount remains in the blood stream and is responsible for the coloring of the urine. If bilirubin does not leave the body, it stays in the blood and can be a sign that the person is suffering from hepatitis or some other blood-related illness.

BLOOD TYPES

There are two kinds of red blood cell types important to the lymphatic system because they involve antigens and antibodies: the **ABO group** and the **Rh factor** (see Table 2.2). The ABO group includes A, B, AB, and O blood types. A and B represent the presence of antigens on the RBC membrane. For example, there are A antigens on the RBCs of the patient with type A blood, while there are B antigens on the RBCs of the patient with type B blood. In someone with type AB blood, both antigens are present in the blood, while type O indicates that the person's blood contains neither A nor B blood.

The plasma of each person's blood contains naturally occurring antibodies for those antigens that are not present in the RBCs. This means that a person with type A blood has anti-B antibodies present in his plasma. In addition, the person with type B blood has anti-A antibodies in his plasma and a type AB blood classification means that the person has neither A nor B antibodies in his plasma. The type O blood patient will have both anti-A and anti-B antibodies (see Figure 2.6).

Blood-typing is extremely important when a blood transfusion is necessary for an operation or other medical procedure. Ideally, a patient should only receive a transfusion of their own blood type, or the procedure will not be successful. For example, if a type A patient needs blood and receives type B blood, then the patient's anti-B antibodies will bind to the donated blood's type B antigens. After the antibodies and the antigens are bound together, they would clump (**agglutination**) and then burst (**hemolysis**), which would defeat the entire purpose of the transfusion. In more serious circumstances, the RBCs that have ruptured would emit free hemoglobin,

TABLE 2.2. Blood Types: The ABO Group

Blood Type	What Are the Antigens on the RBCs?	What Are the Antibodies Present in the Plasma?
A	A	anti-B
B	B	anti-B
AB	both A, B	neither anti-A nor anti-B
O	neither A nor B	both anti-A, anti-B

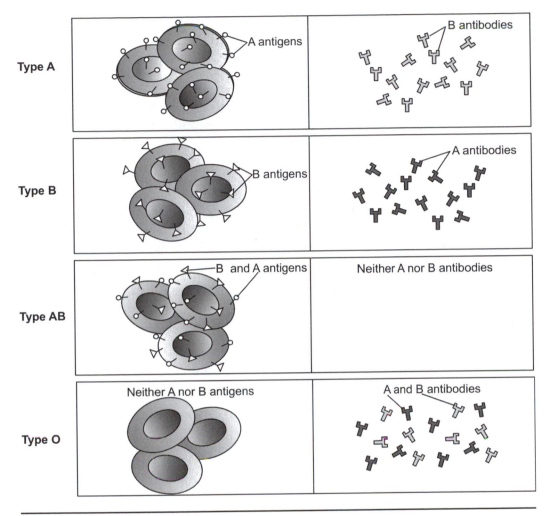

Figure 2.6. Blood types and antibodies.
The column on the left shows red blood cells and the right column shows plasma. The ABO blood types on the left include antigens, while the plasma includes the presence of antibodies.

which would then clog and block the kidney's capillaries, eventually leading to renal failure or even damage. Because the type O patients have neither the A nor the B antigens, they are often considered "universal donors" who will not cause a reaction in the recipient.

Another important characteristic of RBCs is an antigen known as the Rh factor. While people with the Rh factor are considered Rh positive, those without the factor are called Rh negative, and their bodies do not have natural antibodies to this antigen. Therefore, during a blood transfusion, an Rh negative patient should receive Rh negative blood, while the Rh positive patients should receive Rh positive blood. If for some reason an Rh negative patient receives Rh positive blood, the body will perceive the Rh factor as foreign, and therefore will then begin producing antibodies during that initial exposure. While there likely won't be a problem following this initial transfusion, subsequent exposures to Rh positive blood when the anti-Rh factor antibodies are already present could lead to hemolysis and potentially damage the kidneys (see "Newborn Rh Disease").

THYMUS

In addition to the bone marrow, the second primary lymphatic organ is the thymus, which is located under the sternum in an adult (see Figure 2.7). In a fetus and an infant, however, the thymus gland is located below the thyroid gland, which is an endocrine gland below the larynx (the formation

Newborn Rh Disease

When a mother and fetus have opposite Rh factors present in their blood, it can cause erythroblastosis fetalis, or Rh disease of the newborn. While the mother is pregnant, her blood and the fetus's blood do not mix until after the delivery. Shortly following the delivery, the mother releases the afterbirth or placenta, which can cause some of the blood from the fetus to enter the mother's blood circulation.

If the mother is Rh negative and her baby is Rh positive, this initial exposure to the fetal blood will prompt her body to begin producing anti-Rh antibodies. During subsequent pregnancies, these antibodies will pass from the mother, through the placenta and then into the fetal circulation. This will not cause a problem if the fetuses are Rh negative. However, if the fetuses are Rh positive, then the antibodies will destroy the fetal red blood cells (RBCs).

This condition can cause the death of the fetuses, or in less severe cases, the baby will be born anemic due to the loss of the RBCs during the fetal development stages. Doctors will usually begin the baby on a course of gradual transfusion exchanges in order to remove the mother's blood that contains these antibodies, and then replace with the Rh negative blood. However, the baby will continue to produce RBCs with Rh positive factors.

Preventing this disease is actually preferable to the treatment, and a medication known as RhoGam is important in this prevention. Shortly after an Rh negative woman delivers an Rh positive baby, she should be administered RhoGam, which is an anti-Rh antibody which will destroy all fetal RBCs that have found their way into the mother's circulatory system. Ideally, RhoGam will get into the body before the mother's immune system begins producing antibodies. In a matter of months, these RhoGam antibodies break down. Therefore, during subsequent pregnancies, the mother's body will be as if it was never exposed to the Rh positive RBCs.

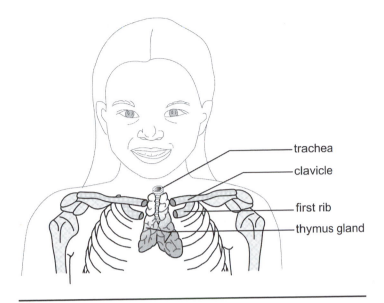

Figure 2.7. Thymus.
The thymus is located below the trachea in a young child.

of the thymus in the fetus is discussed in Chapter 5). As the body develops and grows, the thymus actually shrinks and becomes fat tissue, leaving only a small amount of the thymus in adults. The thymus reaches maximum size during puberty.

The T lymphocytes or T cells (see Chapter 1) which are vital in the body to prepare the immune system to perform its primary duties are produced in the thymus. In fact, the term "T cell" is derived from "thymus-dependent cells." The hormone released by the thymic gland prepares the T cells to recognize antigens and other foreign invaders in the body and subsequently provide immunity.

During the fetal and infant stages, the immune system is immature, which is why babies are more vulnerable to disease and illness in comparison to children and adults. Early in life, the T cells begin (and the lymphocytes perpetuate) this protection as the immune system develops and matures. When a child is 2 years old, his immune system is typically considered mature and fully functional. While infants are routinely given vaccines to boost their immune systems, some vaccines, such as the measles vaccine, are not recommended for children younger than 18 months, because many medical professionals believe that the child's immune system would not be strong enough to respond properly, thus putting the child in danger of not getting the full benefit of the vaccine, or even having a reaction.

3

Secondary Lymphatic Organs: An Examination of the Lymphatic Functions of the Spleen, Tonsils, Adenoid, Peyer's Patches, and Appendix

In addition to the lymph nodes, thymus, and the bone marrow, the lymphatic system functions with help from five other important organs: the *spleen*, *tonsils*, *adenoid*, *Peyer's patches*, and *appendix*. This chapter will describe how these secondary lymphatic organs function and help to protect the body from microbes and other foreign antigens that could lead to illness or cause various kinds of diseases.

SPLEEN

The spleen is protected from harm by the lower rib cage, which encases the organ behind the stomach and is inferior to the diaphragm. While in the fetal stage, the spleen produces RBCs, although this process is taken over by the bone marrow shortly after birth.

There are three primary operations performed by the spleen following birth. One of the spleen's functions is to produce lymphocytes, which then enter the blood and serve as one of the primary tools for the immune system to fight off antigens. Secondly, the spleen contains plasma cells, which produce antibodies that also ward off foreign antigens and microbes. Finally, the spleen also contains macrophages which have the ability to consume, or phagocytize foreign materials floating around in the blood. In addition, the spleen's supply of macrophages serve to destroy old RBCs and produce bilirubin, which is eventually extracted to the liver and excreted from the body as bile.

Because of its two-part composition, this organ is often described as two organs. One portion of the spleen is composed of lymphatic sheaths and germinal centers called *white pulp*. The second portion is known as red pulp, and consists of macrophages. The white pulp's function is considered immune, while the red pulp's function is considered phagocytic. The white pulp is in charge of producing the antibodies, and this region is also where B and T cells, in addition to plasma cells, are produced and mature. The red pulp is kind of a cleaning machine; it removes unwanted matter, such as bacteria. In addition, the red pulp also acts as a reservoir for other lymphatic elements such as white blood cells and platelets.

Doctors do not consider the spleen a "vital organ," because it performs the same functions as some other organs. For instance, the liver and bone marrow can remove RBCs from the circulatory system, and the lymph nodes will produce lymphocytes and monocytes, in addition to destroying pathogens. However, doctors and researchers have found that without a spleen, a person is more vulnerable to certain bacterial infections such as pneumonia and meningitis.

TONSILS, ADENOID, AND THE PEYER'S PATCHES

Tonsils are a type of lymph nodule found in the throat's pharynx. Tonsils are oval-shaped, pink masses of lymphatic tissue. There are three kinds of tonsils named for their location in the pharynx. Along the lateral walls of the pharynx are the palantine tonsils, while the adenoid, or the pharyngeal tonsil, are located on the posterior wall. The lingual tonsils are located on the base of the tongue. The tonsils and adenoid are composed of lymphatic tissue, just like the lymph nodes located in the neck, groin, and armpits.

The adenoid is a single mass of tissue, therefore it is incorrect to refer to it as "adenoids." As explained above, it is located in the upper part of the throat, behind the nose and above a part of the throat called the **uvula**. This area of the throat is called the *nasopharynx*. While the tonsils can be seen simply by opening the mouth wide, the adenoid can only be viewed through the use of special mirrors and instruments that are passed through the nose by a doctor or other medical professional.

The tonsils form a kind of ring of lymphatic tissue around the pharynx. This is a key location because it is near the entrance to breathing passages, in addition to being where food first enters the mouth. Therefore, the tissue can capture germs and pathogens that are coming into the body through food and air, and act as a sort of filter for the lymphatic system. This function is especially important during the initial years of life, but becomes less important as the body and lymphatic system matures. Children who suffer frequent infections of the tonsils (called *tonsillitis*) may have to have their palatine tonsils and their adenoid removed. Some signs of infected tonsils might be noisy breathing, snoring, difficulty swallowing (especially solid foods), and choking or gasping while sleeping. The surgery to remove the tonsils is known as a **tonsillectomy**. It is important to note that when a child has their tonsils removed they do not suffer immunity loss, because the body has redundant systems lined up, such as the other lymph nodules, that will serve the same function if the tonsils are surgically removed (see Chapter 9 for more information about tonsillitis).

In the small intestine, there is an abundance of lymphatic tissue in order to filter out pathogens that might be brought into the body through eating and drinking. One type of lymph nodule grouping located in the small intestine is called Peyer's patches. In the small intestine, the Peyer's patches work to remove pathogens that are invading the body through the digestive system. There are also single lymph nodules, known as solitary lymph nodules, located in the lower part of the small intestine (see the Digestive System volume of this series for more information on how the body breaks down and processes food and other substances).

APPENDIX

A small, blunt-ended tube, the appendix is considered part of the large intestine, even though its walls are rich with lymphatic tissue. Scientists are uncertain about the function of the appendix in the body, and in particular in the immune system. Because it is composed of lymphatic tissue, it could have some sort of immune function, but its purpose is unclear. However, it is recognized that the appendix is not actually part of the digestive tract because of differences in its tissues from the tissues of the small and large intestine. In some cases, fecal matter or waste can become impacted in the appendix, which can cause it to become inflamed, a condition known as **appendicitis**. If this occurs, the appendix will be surgically removed, a process known as an **appendectomy**.

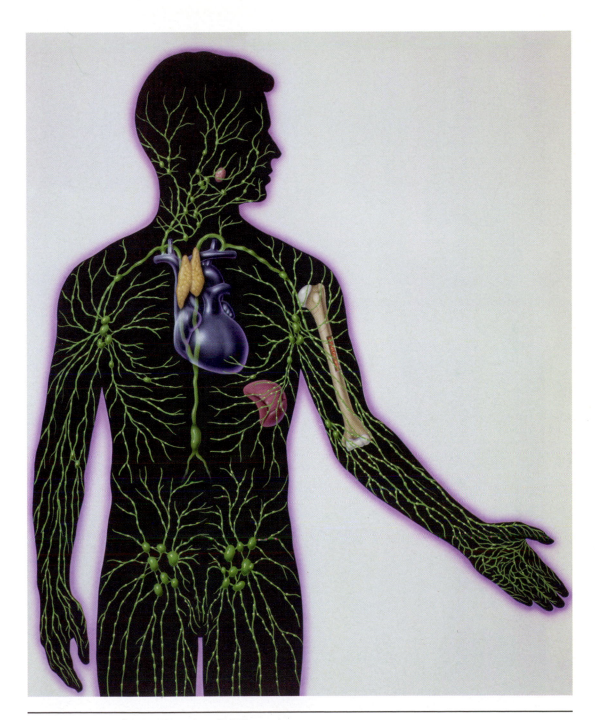

Lymphatic system. © John Karapelou, CMI/Phototake.

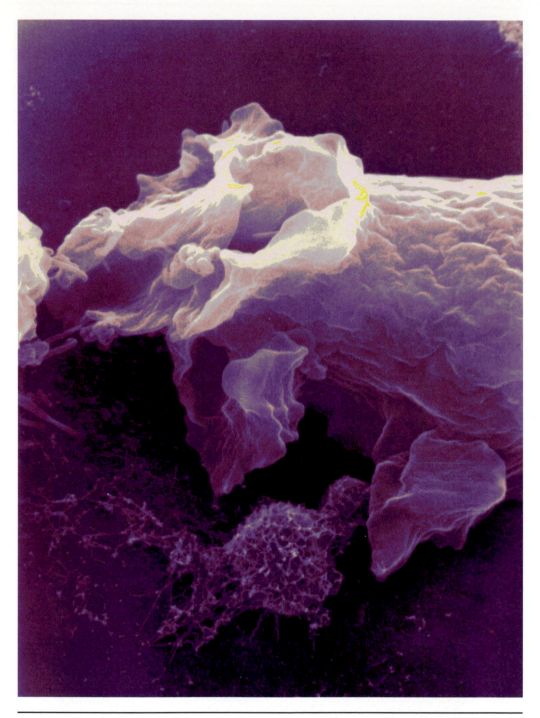

A macrophage attached to an endothelial cell, with lymphocyte attached. © R. Becker/Custom Medical Stock Photo.

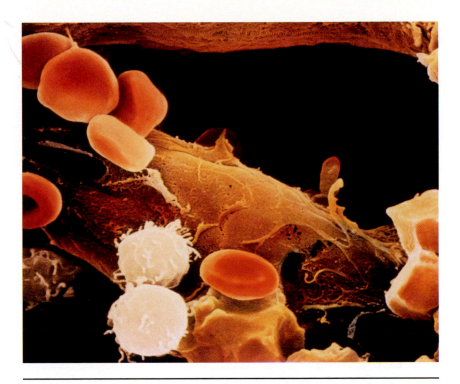

Bone marrow: red and white blood cells. © Prof. P. Motta/Dept. of Anatomy/University, "La Sapienza," Rome/Science Photo Library/Custom Medical Stock Photo.

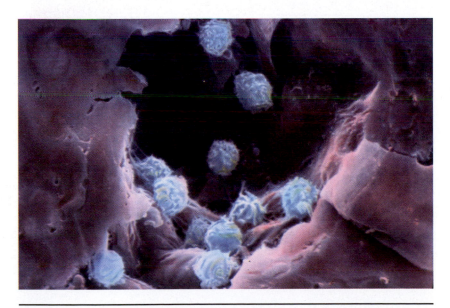

Lymphocytes in a blood vessel. © R. Becker/Custom Medical Stock Photo.

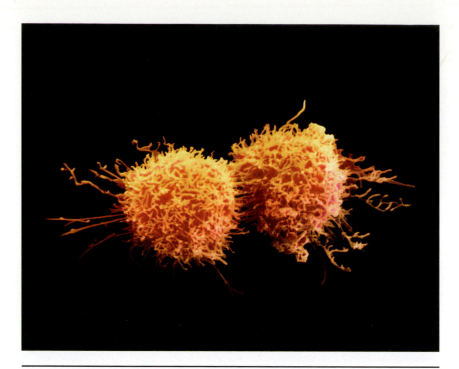

Cancer cells dividing. © David Phillips/Visuals Unlimited.

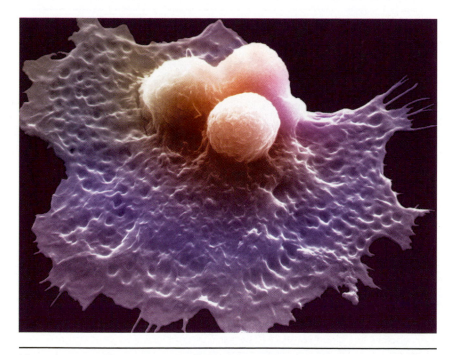

Macrophage engulfing cancer cells. © W. J. Johnson/Visuals Unlimited.

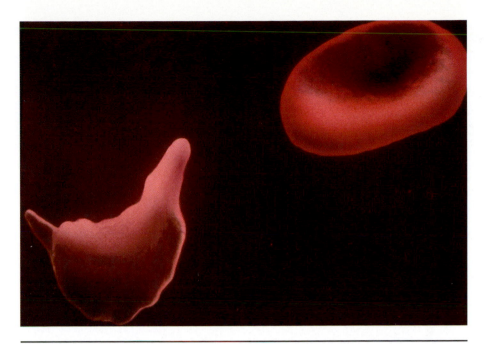

Normal and sickle-cell red blood cells. © R. Roseman/Custom Medical Stock Photo.

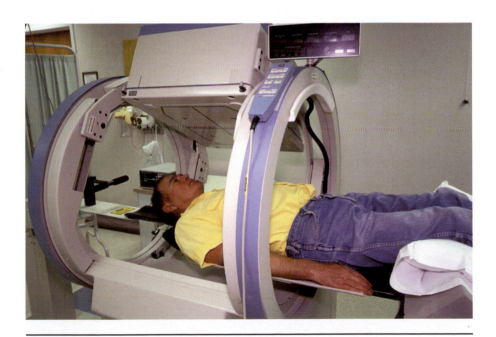

Nuclear medicine scan. © Inga Spence/Visuals Unlimited.

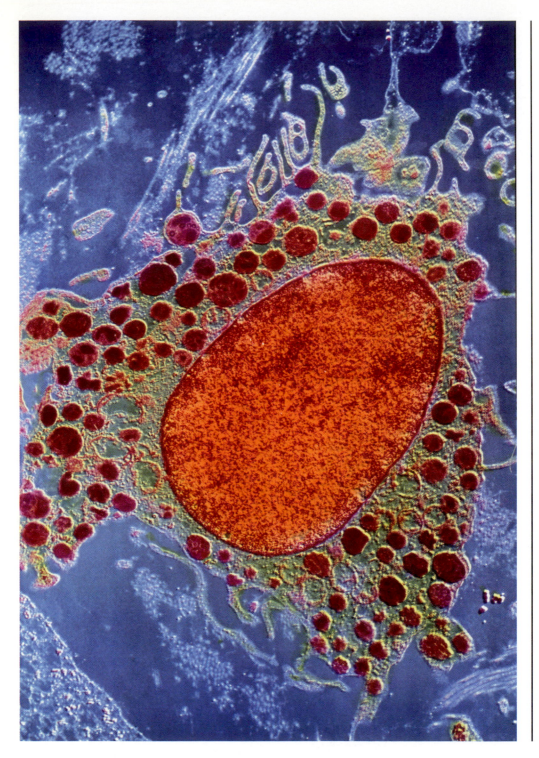

Electron micrograph of a mast cell in an allergic reaction. © Collection CNRI/Phototake.

A doctor checks out a patient's tongue and tonsils. © Custom Medical Stock Photo.

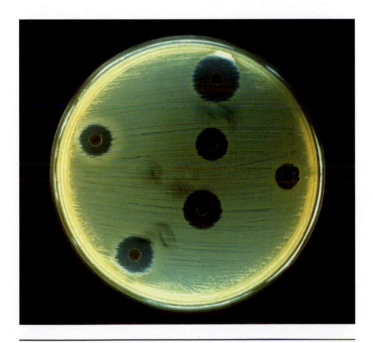

Staphylococcus aureus cultured on an agar plate. © Centers for Disease Control and Prevention.

Food poisoning: a close-up of contaminated jalapeño peppers involved in an outbreak of botulism in Pontiac, Michigan, in 1977. © Centers for Disease Control and Prevention.

4

The Immune Response: Cell-Mediated and Antibody-Mediated Responses

In order to maintain good health and homeostasis, the human body must continually protect itself against harmful disease-causing substances. This chapter will explore how the lymphatic system is able to protect the body by providing **immunity**, the ability to fight off certain infectious substances that can lead to disease and illness. It takes various elements of the lymphatic system, including lymphocytes and antibodies, that provide a strong defense to counter these disease-producing pathogens.

Before each of these lymphatic tools is explained, however, it is important to distinguish between nonspecific and specific defenses. Nonspecific defenses are reactions that involve a variety of pathogens and microbes. These defense reactions occur in the skin, mucous membranes (such as nasal passages), the stomach, and the respiratory tract. For instance, lysozyme is an enzyme produced in the eyes' lacrimal glands, as well as the glands in the mucous membranes of the nose and mouth, that is able to destroy harmful microbes, thus it is considered one of the body's nonspecific defenses. Also, stomach lining is able to secrete an appropriate amount of hydrochloric acid in order to kill harmful microbes present in food, although some food-borne pathogens can still be harmful (see "Food Poisoning"). When one breathes, it's natural to inhale dust particles that then settle in the respiratory tract. Microbes attach themselves to these particles, but thread-like structures called cilia move these microbes up through the respiratory tract to be coughed up or spit out.

Food Poisoning

In the United States, an estimated 76 million people every year suffer—and 5,000 die—from food borne illnesses, according to the U.S. Centers for Disease Control and Prevention. There are more than 250 food-related diseases, most of which are caused by organisms such as bacteria, viruses, and parasites, in addition to natural and manufactured chemicals in food products. These can all contaminate food, which can lead to food poisoning. Some of these illnesses are caused by toxins contained in the contaminant, while other illnesses occur as a result of the body's reaction to the contaminant. Four of the main types of food borne diseases are botulism, campylobacteriosis, E. coli infection, and salmonella.

Botulism is caused by a toxin called *botulinum*, which is produced by the *Clostridium botulinum* bacteria. *C. botulinum* often comes when foods such as asparagus, green beans, and beets are home-canned or preserved. Baked potatoes left sitting out or wrapped in aluminum foil for long periods of time without refrigeration can provide an appropriate site for *C. botulinum* growth. These foods have a low acid content and the bacteria does not need a lot of oxygen to grow, therefore it can spread even in sealed containers. In order to prevent food-borne botulism, experts suggest boiling any home-canned food before eating it to kill bacteria, in addition to refrigerating all food that is not being immediately consumed. When one contracts botulism, a person's nerves are affected, which can lead to paralysis or even respiratory failure in some cases. Medical treatment must be given promptly, and doctors will often introduce an antitoxin into the body's blood stream to counteract the action of the botulism bacteria.

Another common food-borne illness is campylobacteriosis, which is caused by *Campylobacter* bacteria. *C. jejuni* is the type of this bacteria that causes the most cases of this illness, which is the leading cause of bacterial diarrhea in the United States. The most common instances of contamination of *C. jejuni* involve transmission through handling or eating raw or undercooked poultry, drinking nonchlorinated water or raw milk, or handling animal or human feces. Symptoms of campylobacteriosis include severe, often bloody diarrhea, abdominal cramping, persistent nausea, and vomiting. Patients will usually recover with the help of an antibiotic, in addition to drinking plenty of chlorinated water, since diarrhea dehydrates the body.

The third type of food-borne disease is caused by the *Escherichia coli* bacteria, also known as *E. coli*. This bacteria and its toxins are primarily transmitted through undercooked or raw hamburgers; unpasteurized milk, apple juice, or apple cider; and contaminated well water. People have also become contaminated after swimming in water contaminated by sewage. Symptoms of *E. coli* infection are similar to campylobacteriosis: nausea, abdominal cramping, and severe diarrhea. Recovery is usually within a few days with the help of antibiotics.

Salmonellosis, also commonly known as salmonella, can afflict both animals and people, and is often caused by two different bacteria: *Salmonella typhimurium* and *S. enteritidis*. The bacteria have been found in raw poultry, eggs, beef, and sometimes in unwashed fruit. In addition, foods prepared on surfaces that previously contained raw beef or chicken contaminated by the salmonella bacteria can also become infected, which is known as cross-contamination. Salmonella has been known to occur in small outbreaks among a number of people who have eaten at the same restaurant, hospitals, or other institutions that serve mass quantities of food. Symptoms include diarrhea, fever, and abdominal cramps, and can also lead to painful joints and irritated eyes if not treated promptly after infection.

While the body's nonspecific defense system is effective against microbes and some pathogens, it needs help to fight, especially against certain toxins produced by pathogens. Therefore, the body is equipped with a second line of defense, known as the specific defenses. The specific defenses involve the production of antibodies, which serve to inactivate substances called *antigens* (pathogens and their related toxins). Antigens act as chemical markers that identify cells as invasive substances. Human cells have their own antigens, which recognize foreign antigens as dangerous and are subsequently destroyed, thus activating an immune response. Examples of foreign antigens include bacteria, viruses, fungi, protozoa, and **malignant** cells, which are abnormal cells such as those associated with cancer. The response that results from the antibody-antigen reaction is specific. Only a specific antibody can fight off a particular antigen (this relationship will be explored later in this chapter).

Nonspecific and specific immune responses are also referred to as innate and adaptive responses, and both systems work together to identify harmful invaders to the body, and then contain and eliminate them. The innate or nonspecific system is always on alert, and is prepared to react to any and all invaders. While their action is rapid, the innate immune response is also limited, but keeps the harmful pathogens from invading the body to a significant degree. However, the adaptive immune system can come in and is equipped with the powerful, specified tools to completely eliminate the pathogen.

THYMUS GLAND

It is important to emphasize the vital role that the thymus gland plays in immunity. One of the lymphatic system's primary organs (see Chapter 2), scientists and doctors noted in the early years of immune research that children born without this organ could not fight off infection. Research has indicated that the thymus gland is instrumental in structuring and organizing the body's lymphatic system from the fetal years through the initial years after birth. The primary function of the thymus gland is to prepare lymphocytes to participate in the immune response.

LYMPHOCYTES

The lymphatic system's organs—the lymph nodes, thymus gland, spleen, and bone marrow—all contain lymphoid tissue, which is home to two kinds of lymphocytes that each respond to antigens in different ways: T cells and B cells.

In the embryo stage, T cells are produced in the bone marrow and thymus. While passing through the thymus, they mature with the help of the thymic hormones. Scientists believe that the thymus gland alters the lymphocyte's DNA so they become T cells. These cells then travel to the spleen, lymph nodes, and lymph nodules, where they then are produced following birth. After the initial production, the T cells circulate through the body's blood network, and then lodge in the lymph nodes and other lymphoid tissue. T cells are small lymphocytes that attach to antigens. Once attached, the T cells secrete certain enzymes that dissolve the antigen's membrane and digest its contents, which destroys both the antigen and the T cell. These lymphocytes primarily kill antigens produced by fungus cells, viruses, and bacteria that result from slow-developing infections and diseases. T cells are also responsible for the rejection that can result from an organ transplant (see Chapter 9).

The second kind of lymphocyte, the B cell, is produced in the bone marrow during the embryonic stage, although it soon moves into the spleen and lymph nodes and nodules. When B cells come into contact with a specific antigen, they become plasma cells, which produce antibodies that are released into the blood's circulation. Once these specified B cells are produced after coming into contact with a specific antigen, they can remain in the lymph nodes for years, on alert to attack if the antigen is once again introduced into the body.

ANTIBODIES

As stated earlier, antibodies are proteins that are produced in response to foreign antigens. Also called **immunoglobulins** or *gamma globulins*, it is important to note that antibodies in themselves do not destroy foreign antigens. Instead, antibodies attach themselves to foreign antigens, marking these substances so the body knows to destroy them.

As also stated earlier, there is one specific antibody for one specific antigen, which means that if the need occurs, the immune system has the capacity to respond to millions of antigens by producing millions of different antigen-specific antibodies. These millions of antibodies are separated into five classes: IgG, IgA, IgM, IgD, and IgE (see Table 4.1 for a complete description). The structures of each antibody class are depicted in Figure 4.1, with the details of the IgG antibody shown in Figure 4.2. The third drawing, Figure 4.3, shows how the IgG antibody fights bacteria, a virus, and toxins (see "Antibiotics").

THE IMMUNE RESPONSE

One of the main goals of the lymphatic system's immune response is to destroy a harmful pathogen, and the first step to achieve this goal is for the body to recognize the antigen associated with this pathogen as foreign. While both T cells and B cells can provide this recognition function, the

TABLE 4.1. Antibody Classes

Class	Location	Function
IgG	Blood	Provides passive immunity for newborns by crossing the placenta in the mother's womb.
	Extracellular fluid	Provides long-term immunity after a serious illness or following a vaccine.
IgA	External secretions (tears, saliva)	Provides passive immunity, because it is present in breast milk and passed on to newborns; found in all mucous membrane secretions.
IgM	Blood	Initially produced as an infant's immune system matures, also produced following an infant's first infection.
IgD	B lymphocytes	These serve as receptors on B lymphocytes or B cells.
IgE	Mast cells or basophils	Vital in allergic reactions; mast cells release histamine.

Antibiotics

Antibiotic medications are powerful tools because they help stop bacterial infections and other germs that can cause certain illnesses, although they are not effective against infections caused by viruses. Researchers have also found that overusing antibiotics can lead to bacterial resistance. Once a bacteria has become resistant to the antibiotic, the patient is unresponsive to certain kinds of antibiotic medication.

Before prescribing an antibiotic, a physician will determine if the infection is bacteria-based or virus-based. For example, the common cold and flu cannot be cured by antibiotics because they are caused by viruses, as are coughs and bronchitis. If a patient complains of a sore throat, that is also most likely caused by a virus. However, a sore throat can be a sign of strep throat, which is caused by bacteria. Ear and sinus infections can fall into either the bacteria or viral category, and a physician will determine whether antibiotics are needed or whether the body just needs time (sometimes two weeks or more) to get rid of the virus on its own.

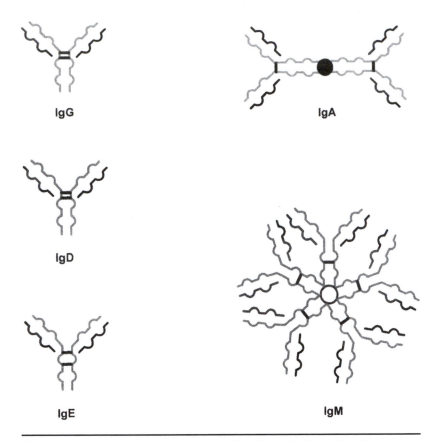

Figure 4.1. The five classes of antibodies—IgG, IgD, IgE, IgA, and IgM.

immune response is more effective if the antigen is dealt with by macrophages and helper T cells, which are a specialized group of T lymphocytes. This foreign antigen is first consumed, or phagocytized, by a macrophage. Parts of this antigen then become attached to the macrophage's cell membrane. Also located on this cell membrane are "self" antigens, which are the safe type of antigens found in other cells throughout the individual. When the helper T cells come into contact with this macrophage, it will detect not only the "self" (and harmless) antigens, but also the foreign antigens. This helper T cell is then on alert and sensitized to this macrophage that contains parts of the foreign antigen.

Cell-Mediated and Humoral Immunity

There are two types of specific (or adaptive) immunity: cell-mediated and humoral. The relationship between the bone marrow, B cells, T cells, and both

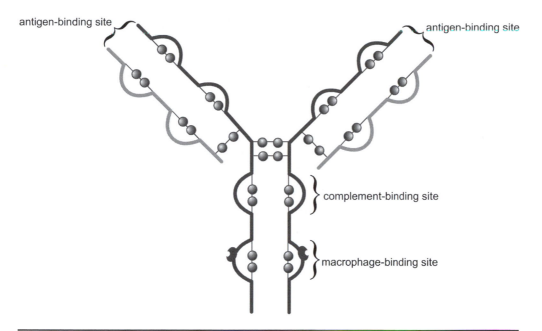

antigen-binding site

antigen-binding site

complement-binding site

macrophage-binding site

Figure 4.2. Detailed depiction of the structure of an IgG molecule.

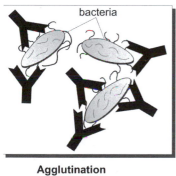

bacteria

Agglutination

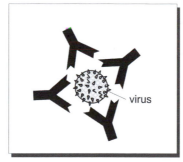

virus

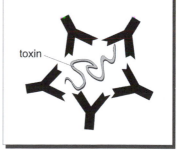

toxin

Neutralization

Figure 4.3. Depiction of antibody behavior.
Bacteria is agglutinated and neutralized by viruses or toxins.

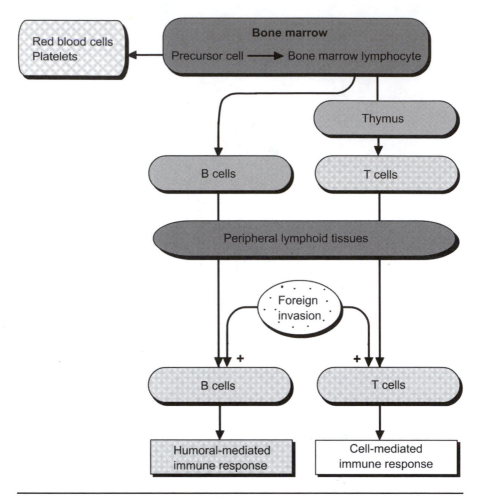

Figure 4.4. Production of B and T cells.
B cells protect the body by initiating a humoral-mediated immune response, while T cells initiate a cell-mediated immune response.

of these immune responses is detailed in Figure 4.4. T cells are responsible for cellular or cell-mediated immunity, which refers to immune response in which macrophages and T cells participate, while humoral immunity involves B cells, T cells, and macrophages. While T cells are associated with cellular immunity, B cells are responsible for humoral immunity, because the B cells are providing protection as they circulate through the blood and tissues of the body. Humoral immunity protects against more acute diseases than are warded off by T cells, such as pneumonia, staphylococcal infection (staph infection), and streptoccoal infection, which is also known as strep throat (see Chapter 9 for more information about these infections).

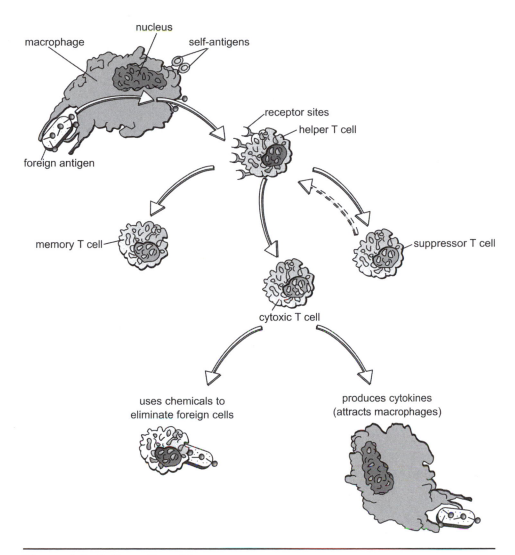

nucleus

macrophage   self-antigens

receptor sites

helper T cell

foreign antigen

memory T cell

suppressor T cell

cytoxic T cell

uses chemicals to
eliminate foreign cells

produces cytokines
(attracts macrophages)

Figure 4.5. Cell-mediated immune response.

Figure 4.5 depicts the cell-mediated immunity process. This process does not produce antibodies, although it is successful against intracellular pathogens, including viruses, fungi, malignant cells, and foreign tissue grafts. The initial step in this process is to activate the T cells, which occurs when the helper T cells and macrophages recognize a foreign antigen. Now recall that these activated T cells are antigen-specific, meaning individual T cells are only successful against certain antigens. But when these T cells are activated, they divide numerous times into two kinds of cells: memory T cells and cytotoxic (killer) T cells. While the memory T cells can always recall a specific antigen and become sensitized upon its presence in the body, the cytotoxic cells are able to eliminate these foreign antigens by destroying their cell membranes. These cells also produce a certain chemical called a **cytokine** that attracts macrophages to the area where an antigen is present, and then activates them to consume or phagocytize the antigen. The effect of these two types of T cells working together ensures that a harmful antigen, such as a virus, is quickly detected, destroyed, and then prevented from reproducing other virus-infected cells in the body.

Another kind of cell, called the *suppressor T-cell*, is also present in this immune system. Once the memory and cytotoxic cells work to eliminate the antigen, the suppressor cells work to stop the immune response once the antigen has been destroyed. However, if the antigen reappears, the memory cells will initiate the immune response.

Unlike the cell-mediated immunity, humoral immunity does result in antibody production (see Figure 4.6). Once again, the initial step in this immune response is also the recognition of the foreign antigen; the helper T cells, B cells, and macrophages are all involved. The helper T cells recognize the antigen, and then alert the B cells, which activate other B cells that might be specialized or specific in combating the antigen. These sensitized B cells go through numerous divisions, which results in the production of two types of cells: memory B cells and plasma B cells. The memory B cells will remember the antigen, while the plasma B cells will produce specific antibodies against this one invading antigen.

After these antibodies are produced, they bond to the antigen, which forms an antigen-antibody complex. This complex is then marked, or undergoes **opsonization**, which means that macrophages or neutrophils will know that this antigen must be phagocytized. The creation of this antigen-antibody complex also begins the process of **complement fixation**. A complement is a family of about twenty different plasma proteins. These proteins circulate through the body's blood network until they are activated, or "fixed," by the formation of the antigen-antibody complex. This fixation process can be complete and thorough, but it can also be partial. Complete fixation is successful if the antigen is cellular, which is often the case with bacterial antigens. In this instance, the complement proteins will bond to the complex and to each other, surrounding the antigen with an enzymatic structure that ultimately inflicts damage on the cells to a destructive degree, thus killing the cell.

But only partial complement fixation takes place if the antigen is not cellular, which would occur if the antigen were a virus. In this instance, only some of the complement proteins bond to the antigen-antibody complex, thus prompting the antigen to go under **chemotaxis**, which is another labeling mechanism that attracts macrophages to phagocytize the antigen.

As stated earlier, once the antigen has been eliminated, the suppressor T

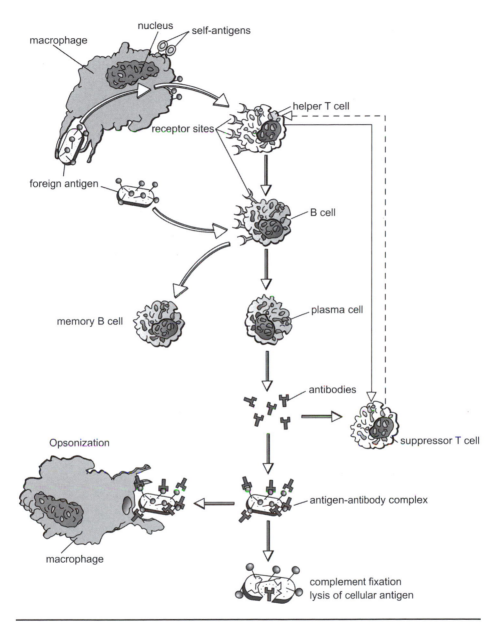

Figure 4.6. Humoral immune response.

cells step in and work to halt the immune response. This is vital to the health of the lymphatic system so that the body doesn't overproduce anti-bodies, which could trigger an autoimmune response (see Chapter 10 for more information on allergies and autoimmune responses).

Innate Immunity

Two important types of innate, or nonspecific, defenses are the inflam-mation response and the release of interferons (see Chapter 1). The inflam-mation of body tissue is a nonspecific response to tissue injury or pathogen invasion. The inflammatory response has three primary goals: the isolation and elimination of the harmful pathogens, the removal of debris from the injury site, and the preparation of the injury site for healing and repair.

The inflammation process is depicted in Figure 4.7. When a pathogen, such as bacteria, enters the skin by breaking through the external skin wall, the macrophages present in that region of the skin promptly descend and go to work phagocytizing the microbes. As the initial line of defense, the macrophages fight infection to some extent, although they are not able to shoulder the work on their own. In fact, they are relatively stationary by na-ture, although they can travel to other sites near that initial region of inva-sion if necessary. As the macrophages go to work on the microbes, mast cells in the area of tissue damage release histamine, which prompts **vasodilation**. As a result of vasodilation, blood vessels expand, delivering an increased amount of blood, phagocytic leukocytes, and plasma proteins to the injury site. In addition, the release of histamine causes capillaries to become more permeable, which causes a capillary's pores to enlarge. When this occurs, plasma proteins that are normally trapped in the capillaries are able to es-cape and travel to the inflamed tissue.

The arrival of the leukocytes, plasma proteins, along with the accompa-nying increased amounts of blood, cause fluid to build up in the injured tis-sue area. This leads to swelling, one characteristic of inflammation, in addition to redness and heat, which are due to the increased flow of warm, arterial blood to the region. These substances also sensitize the afferent neu-rons in the area, which causes the feelings of pain and tenderness (see "Anti-Inflammatory Drugs").

In addition to inflammation, interferons are another nonspecific immune defense that are important against viral infections. Interferons are not one single chemical or molecule, but rather a family of proteins that interfere, albeit briefly, with a virus's ability to reproduce in other cells. These pro-teins derive their name from the ability to interfere with viral replication.

When a virus invades a cell, the cell's genetic components prompt it to produce an interferon, which is then secreted into the extracellular fluid (see Figure 4.8). In this way, the interferon is able to warn nearby healthy

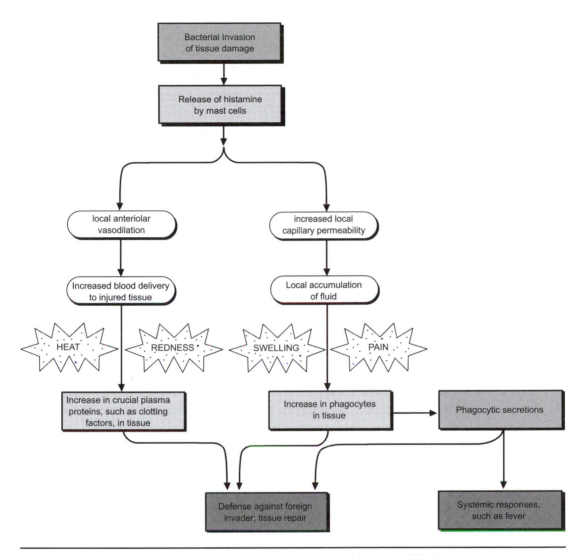

Figure 4.7. The production of inflammation as part of the body's immune response.

cells of the viral presence, thus helping them to prepare for resistance. Once the interferon is in the extracellular fluid, it attaches itself to receptors present on the plasma membranes of these neighboring cells, and even distant cells. While it is important to note that interferons themselves do not have antiviral capabilities, these proteins prompt potential host cells to begin production of virus-blocking enzymes, which can break down the virus's genetic components (including protein synthesis) that are vital for replication. These antiviral enzymes are dormant unless a virus invades their host cells, in which case the enzymes detect the virus's genetic material (nucleic acid)

Anti-Inflammatory Drugs

Two of the most effective drugs used to suppress inflammation are related to *salictylates* (which includes aspirin) and **glucocorticoids** (which are medications similar to **cortisol**, a hormone secreted by the **adrenal gland**). Salictylates work to diminish the release of histamine from the mast cells, which reduces inflammation and the accompanying swelling, redness, and pain. These drugs work to block two enzymes known as cyclooxygenase-1 and cyclooxygenase-2, which stimulate the production of **prostaglandin** hormones. Prostaglandin hormones cause tissue inflammation, **fever**, and pain.

In addition to suppressing inflammation, glucocorticoids also work to destroy lymphocytes and reduce the production of antibodies. Glucocorticoids are used to medicate undesirable immune responses, such as allergic reactions (see Chapter 10) including poison ivy and insect bites, in addition to arthritis-associated inflammation. However, because glucocorticoids suppress immune response, they also increase the body's vulnerability to infection. Therefore, it is important that these medications be used only when necessary and under a doctor's supervision.

and go to work. However, this activation can only occur during a limited amount of time, and therefore is only a temporary defense against viral infections. After this time, specified immune responses are needed, such as those that relate to antibody production.

In addition to their power against viral replication, interferons also help to boost other immune-related mechanisms, including enhancing the phagocytic behavior of macrophages and the production of antibodies. Because they have been shown to slow cell division and tumor growth, interferons also have anticancer benefits as well, although the extent of these benefits is still being researched (see "Interferons: Miracle Cure?").

GENETIC AND ACQUIRED IMMUNITY

The human body's immune system is divided into two categories: **genetic immunity** and **acquired immunity**. Genetic immunity is determined by DNA and therefore does not involve antibodies; rather it is the immune system that we are born with as part of our genetic composition. This is specific to a species, however, which means that some species have immunity to pathogens that are harmful to other species. For example, while the measles virus is harmful to humans, it does not affect dogs and cats because of their genetic immunity to the pathogens associated with the virus. Certain viruses that are dangerous to plants are harmless to humans because our genetic makeup protects our cells and tissues from these pathogens (see Table 4.2).

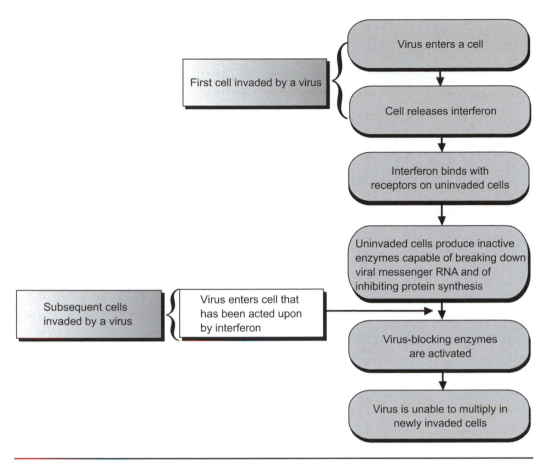

Figure 4.8. Interferon prevents the replication of virus-infected cells.

Interferons: Miracle Cure?

When interferons' potential benefits first gained attention in the mid-1950s, they were hailed as a powerful therapeutic tool against illnesses ranging in severity from the common cold to cancer. However, scientists and researchers soon found that it was impossible to collect a sufficient quantity of human interferons in order to determine their medicinal potential. In the 1980s, advancements in genetic research allowed scientists to produce interferons in a laboratory setting, giving scientists the quantities needed for rigorous scientific study. Unfortunately, many initial studies involving interferons' effects against cancerous tumors and other virus-related infections were not encouraging, and the hype surrounding interferons' potential as a miracle drug began to fade. However, interferons have been shown to be effective against some forms of cancer, including hairy cell leukemia (see Chapter 8) and the AIDS-related cancer Kaposi's sarcoma. In addition, interferons have been shown to be effective in treating the nervous system disorder multiple sclerosis.

TABLE 4.2. Genetic and Acquired Immunity

Type of Immunity	Definition	Examples
Genetic	Part of the human body's genetic makeup (DNA); does not involve antibodies; different from other species.	
Acquired	Involves antibody production.	
	Passive (Natural)	Transmission of antibodies through placenta from mother to fetus; antibodies from other resources; antibody transmission through breast milk from mother to baby.
	Passive (Artificial)	Antibody injection in the form of gamma globulins or immune globulins after prior exposure.
	Active (Natural)	Production of own antibodies; disease recovery through antibody and memory cell production.
	Active (Artificial)	Antibody and memory cell production as a result of a vaccine.

Acquired immunity involves antibody production. Within this category, there are two subcategories: passive and active immunity. While active or "self-generated" immunity involves the production of antibodies, passive or "borrowed" immunity is when antibodies are acquired through the direct transfer of antibodies produced by another person or animal. An example of this transfer includes the IgG antibody movement from mother to fetus through the womb's placenta. Another example includes breast-feeding, when a mother's milk is enriched with IgA antibodies. This is important because even though these acquired antibodies are broken down within a month, they provide that initial protection against infections while the newborn begins to develop his own immune response system, because the production of antibodies does not happen until at least one month after birth (see Table 4.3 for a breakdown of passive and active immunity).

In some cases, passive immunity is used in response to a lethal toxin or dangerous viral-related infectious agent, such as when someone has been bitten by a dog afflicted with the rabies virus or a snake with poisonous venom. When this occurs, the injected antibodies have been harvested from another, often nonhuman source that has produced antibodies following ex-

TABLE 4.3. Active and Passive Immunity

	Active Immunity	Passive Immunity
Antigen exposure	The immune system is exposed to the antigen either naturally as a result of an infection, or artificially as a result of a vaccination.	No antigen exposure is required.
Antibody source	Antibodies are produced by the body's immune system upon exposure to an antigen.	Preformed antibodies are borrowed from another source and introduced into the immune system.
Administration of antibodies	Injection or attenuated antigen is needed for artificial active immunity.	For artificial passive immunity, patient is injected with borrowed antibodies.
Develop time for resistance	Antibody production can occur in several weeks or days.	Antibody production is immediate.
Resistance time	Sometimes lifelong.	Only a few weeks.

posure to the antigen in an attenuated form, which means a less potent or virulent form of the antigen. Some of the most common animals used for passive immunity procedures are horses and sheep. Because the antibody injection (or serum) is a foreign substance, the body's immune system might launch an attack in response, which would result in an allergic reaction.

VACCINES AND IMMUNITY

While active immunity is the production of one's own antibodies, it may be done naturally, or even stimulated through artificial means. Once a person has recovered from a disease, such as chickenpox, his body now has specific antibodies and memory cells that will activate if that disease's pathogens enter the body again. This is called *naturally acquired active immunity*. Vaccines help bring about artificially acquired active immunity, because the mechanism stimulates the production of memory cells in addition to antibodies against a disease.

The vaccine mechanism was invented by Edward Jenner, a British physician in 1797, when he found that patients could become immune to the devastating effects of the smallpox disease (which had a mortality rate of 40 percent) through exposing or inoculating patients with small amounts of the cowpox disease, which is a weaker form of the smallpox disease (see Chapters 6 and 7 for more information on the discovery of vaccines).

The goal of vaccination, which begins soon after birth for infants, is to equip the body with active immunity by creating a memory system that calls on B and T cells to defend the body when recognizable disease-causing agents enter the blood and lymphatic system. Usually injected through a needle, a vaccine contains a weakened, or attenuated, form of a specific disease-causing germ. In some cases, the vaccine will contain an inactivated form of the germ, which will produce the toxins associated with the disease upon entering the bloodstream. Once in the body, the immune system produces antibodies against these germs. Because these germs are weakened or dead, they are often not strong enough to make the patient sick, but just strong enough to stimulate the immune system to produce antibodies. It's important to note, however, that just like any medical procedure, there are risks associated with vaccines. Some people have an allergic or other adverse reaction to certain vaccines, although the chances are minimal. Once these antibodies are floating around in the body, memory cells are formed. Both antibodies and these memory cells are then on alert in case the disease-related germ or pathogen invades the body at a later date.

Vaccines come in different forms. Some are used in combinations, such as the DTP (**diphtheria**, tetanus, pertussis) and the MMR (measles, mumps, rubella) vaccines that children receive during their early years to boost their immune systems. Vaccines can be produced from three types of microbe-related materials: inactivated (killed), attenuated (live), and synthetic (produced in a laboratory). The first type, inactivated vaccines, are produced by killing the disease-causing microbe through chemical means, making them stable. Most inactivated vaccines stimulate a rather weak immune response in patients, therefore they must be given several times. Examples of inactivated vaccines include those for cholera and hepatitis A. Live and synthetic vaccines are produced in a laboratory setting in order to eliminate its viral and disease-causing characteristics. Unlike inactivated forms, these vaccines produce both cell-mediated and antibody-mediated immunity, and usually only require one dose. Examples of these vaccines include those for yellow fever, measles, rubella, and mumps.

Another kind of vaccine formulation is a **toxoid**. A toxoid contains an inactivated form of a toxin, which is the harmful substance produced by a microbe. In general, microbes are not dangerous or disease-causing, although some toxins emitted by microbes can cause illness. For example, in a normal environmental setting with plenty of oxygen, the bacterium associated with tetanus is harmless. But when this bacterium is in an environment without oxygen, the bacterium produces a poisonous toxin. These potent toxins are treated with materials, such as a sterile water and formaldehyde solution called *formalin*, which inactivates these toxins. Diphtheria and tetanus vaccines use toxoids.

AGING AND THE IMMUNE SYSTEM

As a person ages, it becomes harder for the body to fight off infection and diseases because of the diminishing ability to mount a defense against harmful pathogens. The human body's immune system does not function as efficiently in older adults as it does in younger adults and teenagers.

The progressive decrease of the immune system's ability to protect the body is known as immune senescence, which leads to inappropriate and in some cases, detrimental responses when the body is trying to fight off disease and infection. As explained earlier, there are two complementary forms of immunity that protect the human body: natural, or innate, and adaptive, or acquired, immunity. This means that while natural immunity rapidly responds against invading pathogens, this response is incomplete until the slower but more definitive adaptive immune response develops to provide comprehensive protection. It is important to note that natural immunity is characterized by a relatively rigid structure, while adaptive immunity is more flexible as the name suggests, and is supported by both T and B lymphocytes.

Each type of immunity experiences different changes throughout the aging process. In order to understand how natural immunity responds to aging, each of the system's elements—dendritic cells, macrophages, natural killers cells, and the complement system—should be given a close examination. Dendritic cells are responsible for processing antigens, one of the important initial stages for initiating an immune response. These cells prepare the antigens to be presented to CD4+ lymphocytes. Even though the effect of aging on dendritic cells has not be thoroughly studied, scientists believe that in general, the elderly do not have as many of these cells as younger adults and teenagers. However, those dendritic cells that are present in elderly adults are efficient and are able to function at full capacity, helping to induce an immune response in addition to aiding in the proliferation of the T lymphocytes.

The second element of natural immunity, macrophages, are also involved in those initial steps in inducing an immune response. In the elderly, studies have found that macrophages enable a normal T lymphocyte response to occur in the body. However, the speed that macrophages process antigens decreases with age. In addition, macrophages are not as powerful against tumor cells in the elderly, which may explain why cancer rates increase among the older populations.

The activity rate of natural killer cells, the third element of natural immunity, actually increases with age, or remains unchanged. In contrast, the fourth natural immunity element, the complement system, diminishes in activity in older patients. This means that in the case of a bacterial infection, complement levels will increase significantly in younger patients compared to elderly patients.

In terms of adaptive or acquired immunity, there are also four elements that are affected by age: thymus activity, T lymphocyte function, B lymphocyte function, and autoimmune reactivity. As stated earlier, the thymus is where T lymphocytes produced in the bone marrow mature and become functional and effective. Soon after birth, the thymus begins regressing and continues at a constant rate, almost as if undergoing accelerated premature aging. This means that the T lymphocyte–mediated response begins to decrease soon after birth. However, as the thymus regresses, peripheral lymphoid organs compensate for the thymus, eventually rendering it unnecessary. In fact, studies of mature animal models with thymic transplants have not shown to significantly improve immune response. There are several hormones produced by the thymus that influence the differentiation activity of T lymphocytes, and many scientists believe these could have an effect on mature B and T lymphocytes. These hormones have been found to decrease in concentration in elderly adults, although its not clear how this effects the performance of the immune response.

The second element of adaptive immunity—T lymphocyte function—appears to be the most affected by age. Studies have shown that the ability of T lymphocytes to respond to antigens decreases significantly with the maturation of the immune system. Throughout the aging process, the production of T lymphocytes decreases, although there is an increase in memory T lymphocytes that are already familiar with antigens. In comparison, "naïve" T lymphocytes decrease with age. This shift to a greater amount of memory T lymphocytes begins early in a patient's life, and scientists believe that the progressive expansion of these cells over a lifetime contributes to some aspect of immune senescence. Elderly patients might also have T lymphocytes with aberrant or abnormal characteristics. In addition, biochemical lesions have been found to accumulate along pathways where nervous system–initiated signal transmissions occur, which can stimulate genes involved in T lymphocyte activity. This can interrupt signal transmission between the nervous and immune system throughout the body.

Another element of adaptive immunity, the B lymphocyte function, is responsible for making antibodies. Scientists have found that as a person ages, their B lymphocytes decrease, which means that their antibody production also decreases. In particular, antigen-responsive B lymphocytes that circulate through the body decrease in number.

In terms of adaptive autoimmune reactivity, the number of autoreactive T and B lymphocytes appears to increase with age. This is the opposite of the diminishing effect that aging has on the immune system's ability to fight off exogenous antigens. The number of autoreactive T and B lymphocytes not only increases with age, but the frequency of autoantibodies that protect the body against organ-specific and non-organ-specific antigens increases as well.

The Development of the Lymphatic System

The development of the human body from a single-celled **zygote** into an organism with in excess of 100 trillion cells is an amazing biological phenomenon. The study of the developmental pathways following fertilization is called *embryology*. The embryology of the lymphatic system is complex, but includes two major areas. First, there is the embryological development of the lymphatic vessels and primary and secondary lymph organs. The second major area of immune system development during the embryonic stage involves the formation of the white blood cells and lymphocytes. These cells are derived from the same types of tissues as the cells of the circulatory system. In fact, there are many similarities between the early development of the lymphatic cells and the circulatory cells. The term *hematopoietic* is sometimes used to describe all aspects of circulation in the body. This chapter will focus on the embryological development of the lymphatic system from conception until just prior to birth.

FORMATION OF GERM LAYERS

Life starts as a single cell called the zygote, and then progresses through a series of cell divisions forming a hollow mass of cells, called a **blastula** (see Figure 5.1). During these first few divisions the cells are undifferentiated, meaning that they do not yet have specific functions, and they are not increasing in mass. In fact, during the first week of life, the embryo remains about the same size as the ovum; only the number of cells increase.

As the blastula forms, the cells begin to form two distinct structures (see

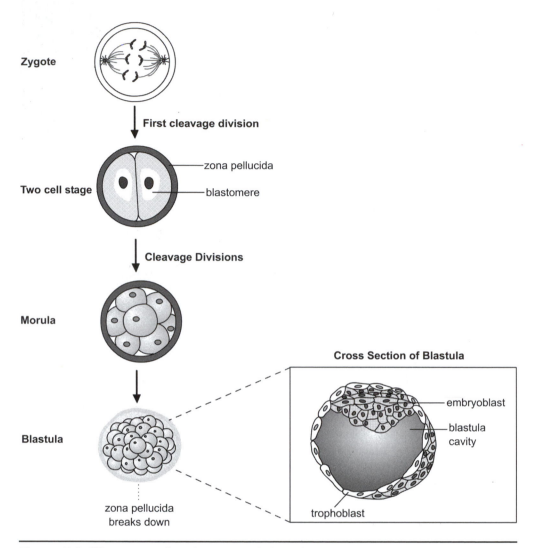

Figure 5.1. The stage of embryogenesis in animal cells.

Figure 5.1). One group of cells, called the *trophoblasts*, retains the original ball shape of the embryo. A second group of cells begins to form the inner cell mass (ICM), which is also sometimes called the *embryoblast*. Embryoblasts are responsible for the formation of the body tissues. Shortly after the formation of the ICM, the embryo enters into the uterus and implants itself into the uterine wall; this is the work of the trophoblast cells. Because the cells of the embryo are different genetically from the cells of the mother's uterus, the embryonic cells should cause an immunological reaction by the mother's body. However, this does not happen, although the reason is not yet completely understood.

TABLE 5.1. General Functions of the Major Tissue Types

Tissue	Derived from	General Function	Examples
Nerve	Ectoderm	Forms neurons and associated cells of the nervous system	Cells of the central and peripheral nervous systems
Connective	Mesoderm	Physical connection of organs and tissues; cellular components of blood	Lymphocytes, blood cells, bone, cartilage
Muscle	Mesoderm	Movement of body; contraction of organs and heart	Smooth, skeletal, and cardiac muscles
Epithelial	Endoderm Mesoderm Ectoderm	Forms sheets of cells that represent the protective coverings	Linings of the respiratory and digestive systems

Over the next two to three weeks, the trophoblast cells are responsible for the formation of the majority of the support structures necessary to link the mother's body with the developing embryo. However, the purpose of this chapter is to focus on the aspects of embryological development directly associated with the formation of the lymphatic and immune systems. These systems, as are all body systems, are derived from the cells of the ICM. (Readers who are interested in the specifics of embryonic development, including mother-embryo interactions, should consult the Reproductive System volume of this series and the references in the Bibliography section of this volume for additional information.)

By the third week of development, the cells of the ICM have started to fold and form what are called *germ layers*. There are three germ layers: the endoderm, mesoderm, and ectoderm, which are named for their initial location in the embryo. These germ layers are responsible for the formation of the four primary types of tissue: nerve tissue, connective tissue, muscle tissue, and epithelial tissue (see Table 5.1 for a general summary of these tissue types). The cells and organs of the lymphatic system are derived from a combination of the connective and epithelial tissues.

FORMATION OF CONNECTIVE AND EPITHELIAL TISSUES

Once the germ layers are formed, they begin to assume a folded shape that is important in the development of most of the structures of the body.

For example, certain folds of the endoderm are responsible for the formation of the gastrointestinal tract, while others form the central nervous system. The cells associated with immune response, namely the lymphocytes, are formed by the embryonic structures that give rise to connective tissue, while the lymphatic organs are comprised of connective tissue with an epithelial lining.

Epithelial cells may be formed from any of the three germ layers. In lymphatic system development, the epithelial linings of the thymus and tonsils are derived from the endoderm. After the endoderm develops, it begins to form a tube-shaped structure that runs the length of the developing embryo. From this tube, small buds of epithelial tissue form, which eventually will provide the coverings for organs such as the liver, pancreas, and respiratory tract. In the head area of the embryo, a group of epithelial buds, called the *pharyngeal buds*, give rise to the epithelial linings of the tonsils and thymus.

All connective tissue is formed from the mesoderm layer, and this connective tissue is responsible for the majority of the organs in the body. One exception to this is the thymus. The thymus is actually derived from two different sets of epithelial tissue. In addition to the epithelial cells from the endoderm tube, a second set of epithelial cells are formed from the mesoderm layer. These cells will form the bulk of the cells within the thymus.

By the fourth week of development, the mesoderm is organized into four distinct regions: axial, paraxial, intermediate, and lateral plate. The axial mesoderm is responsible for formation of the **notochord**. The paraxial mesoderm divides into specialized sections of compacted cells called somites. There can be between forty-two and forty-four pairs of somites in the developing embryo. In most cases, it is the somites that are responsible for the formation of the bones, connective tissue of the skin, and the skeletal muscles. The intermediate mesoderm is involved in the formation of the urinary and reproductive systems.

The final category of mesoderm, the lateral plate mesoderm, is subdivided into two additional classes: the somatic mesoderm and the splanchnic mesoderm. The somatic region is primarily associated with the formation of smooth muscle and connective tissue of the arms, legs, wall of the body, and the head region. Finally, it is the splanchnic mesoderm that gives rise to much of the connective tissue associated with the circulatory system, including the lymphatic system.

ORIGIN OF THE LYMPHOCYTES

Lymphocytes (the precursors of the B and T cells; see Chapter 1), are formed from the same cells that give rise to the red blood cells and other cells of the circulatory system. These cells are called the *hematopoietic cells,* and they are formed from stem cells. Stem cells are undifferentiated cells that can form a variety of new cell types. The process by which stem cells become new lymphocytes (and other cells) is called *hematopoiesis.*

The hematopoietic stem cells may enter into one of two developmental pathways. The first is the myeloid line, which is responsible for the eventual formation of red blood cells (erythrocytes), monocytes (macrophages), neutrophils, basophils, and eosinophils (the function of the macrophages, platelets, neutrophils, basophils, and eosinophils is covered in more detail in Chapter 1). The second pathway of development is called the *lymphoid line*, and as the name implies, it is responsible for the formation of the lymphocytes, or B and T cells. These cells are formed from lymphoid precursors, but mature in the bone marrow or thymus, respectively (see Chapter 1).

The site of hematopoiesis varies throughout the development of the embryo. Around the third week, the red blood cells begin to be formed by the cells of the yolk sac, but the yolk sac only retains this activity for a brief period of time before the task is transferred to the cells forming the liver. By week five, the liver has become a major site of hematopoietic activity, and it retains this ability until just prior to birth. However, at about the same time, the spleen becomes active as well. While all cell types may be formed in the spleen, it is a major site of red blood cell production until the bone marrow matures.

FORMATION OF THE PRIMARY AND SECONDARY ORGANS

As the primary and secondary lymphatic organs and glands (see Chapters 2 and 3, respectively) begin to develop during embryogenesis, they are colonized by lymphocytes coming from the temporary hematopoietic organs listed in the previous section. However, by the sixth month, these systems are largely being replaced by the bone marrow, although some of the organs retain their function until birth.

By the second month of life, the organs that will eventually comprise the lymphatic system are becoming functional. The lymph nodes, including the tonsils, are formed from sacs of connective tissue. In the case of the tonsils, the lymph nodes are partially protected by epithelial cells from the endoderm.

As mentioned previously, the thymus is relatively unique in that it is primarily comprised of endoderm epithelial tissue. This organ starts to form around the sixth week and is completely functional by around the fourth month of development. Thus, by the sixth month of development the lymphatic system is mostly operational, and the cells of the immune system are maturing in the bone marrow.

6

History of Discovery from the Ancients to the Nineteenth Century

Unlike many of the other major systems in the body, such as the respiratory and digestive systems, the existence of the lymphatic circulation and immune functions in the body has not been well understood until modern times. Scientists and medical professionals are still unraveling the mysteries of the immune system. Still, before the twentieth century there were a number of experiments and investigations that, while not providing a comprehensive understanding of the role of this body system in human physiology, did contribute important clues and information that served as the foundation for later studies in the twentieth and twenty-first centuries.

The study of the history of the lymphatic system actually follows two major tracts. First, there are the discoveries that led to the understanding that the lymphatic system represents a separate form of circulation in the human body. Although investigations on the circulation of lymph began in ancient times, it was not until the seventeenth century that plausible explanations were developed on the physiology of lymph circulation. The second major area of study is the origin and function of the chemicals and cells that give rise to the immune response. Studies in this second area did not effectively begin until the nineteenth century, and coincided with the scientific studies of immunization in western Europe (see "Chronology of Advances"). However, it was not until the twentieth century that the relationship between these cells and immunity was established (see Chapter 7).

Chronology of Advances in the Study of the Lymphatic System from the Sixteenth to Nineteenth Centuries

1561–1564	Both Falloppio and Eustachio mention the presence of lymphatic vessels in the body, although neither provides a description of their function.
1622	Aselli makes the first detailed description of the lacteals of the digestive tract.
1647–1651	Pecquet discovers that the lymphatic system connects to the circulatory system at the thoracic duct.
1652	Bartholin publishes *De Lacteis Thoracsis*, in which the name *lymphatic system* is first used.
1653	Rudbeck publishes a comprehensive study of the lymphatic system of animals; he supports the idea that the lymphatic system represents a separate circulatory system in the body.
1752–1754	Munro and Hunter independently describe the role of the lymphatic system in returning fluid from the tissues of the body.
1796	Jenner conducts the first intentional vaccination of smallpox in Western medicine.
1843	Addison describes the movement of leukocytes during inflammation.
1862	Pasteur develops the germ theory of disease, which plays an important role in the development of vaccines.
1881	Pasteur develops a vaccine against anthrax.
1883	Metchnikoff describes the phagocytic nature of white blood cells.
1885	Pasteur develops the rabies vaccine, the first vaccine against a virus.
1890	von Behring develops the use of antitoxins as a method of vaccination.

THE ANCIENTS

The ancient Egyptians were one of the first major cultures to keep records of their study of human anatomy, primarily due to the elaborate procedures for the preserving of the human body in preparation for the afterlife. There can be little doubt that Egyptian men of science noticed lymphatic structures such as the thymus, appendix, and spleen, as well as some of the more prominent lymphatic ducts, but there is little evidence that they understood the purpose of these organs.

There is also evidence that the ancient civilizations of the Persians, Chinese, and Indians experimented with the protection of their population from specific diseases by exposing individuals to a related form of the disease that had a lower level of lethality. The ancient Chinese exposed people to the dried scabs of smallpox victims to protect them from more virulent forms of the disease. These were the first attempts at developing vaccinations. However, these cultures were unaware as to the agents causing the disease, and these practices had relatively low levels of effectiveness. The development of modern vaccinations would not start until the eighteenth century.

The ancient Greek scientists built extensively on the knowledge of the ancient Egyptians, but actually made little progress in recognizing that the lymphatic system even represented a body system. Greek scientists such as Herophilus (ca. 300 BCE) and Erasistratus (ca. 280 BCE), and even Aristotle (384–322 BCE), noted the presence of lymphatic ducts in the small intestine (also called the *lacteals*; see Chapter 3). These structures tend to swell after a meal, making them easy to spot along the length of the intestinal tract. These individuals believed that the other lymphatic vessels of the body were the same as the veins of the circulatory system.

The Greek scientist Galen (131–201) is responsible for developing many of the longest lasting theories of human anatomy. His views on digestion, respiration, and circulation, which were rarely correct, dominated medical studies until the sixteenth century. Galen did not consider the lymphatic vessels to be the same as veins, which he also did not believe were involved in the process of blood circulation (see the Circulatory System volume of this series for more information). However, he did not offer an alternative suggestion as to the role of the lymphatic vessels in the human body. So, for the next thousand years, there were few attempts to understand this system.

Galen's authority in physiology and anatomy largely persisted in Western science until the Renaissance. During the Renaissance, there was a renewed in-

PRIMA. SECVNDA. TERTIA. QVARTA.

QVATVOR DECIMAENONAE FIGVRAE TABELLARVM earundemᶐ characterum Index.

Andreas Vesalius, "Four different views of the spleen." From *De humani corporis fabrica*, 1543. © National Library of Medicine.

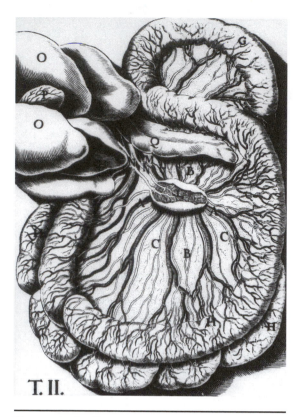

Gaspare Aselli, "The lymphatic system." Woodcut from *De lactibus sive lacteis venis*, 1640. © National Library of Medicine.

terest in the sciences, including the study of the human body. This was primarily a descriptive phase in the history of science, when anatomists such as Leonardo da Vinci (1452–1519) went to great lengths to describe the physical characteristics of the human body. The structures of the lymphatic system did not go unnoticed during this time. The great Renaissance scientist Vesalius (1514–1564), an individual who challenged many of the remaining misconceptions of Galen's that persisted into the sixteenth century, noted the presence of the lymph ducts, but incorrectly considered them to be part of the circulatory system. Later in that century, investigators such as Bartolommeo Eustachio (1520–1574), pioneer in the study of the human ear, and Gabriel Falloppio (1523–1562), one of the first scientists to describe the details of the female reproductive system, both again mentioned the presence of the lymphatic ducts, but, as was previously the case, failed to speculate on the function of these vessels.

THE SEVENTEENTH CENTURY AND THE SCIENTIFIC REVOLUTION

The Scientific Revolution is frequently identified as the time when the **scientific method**, and its emphasis on experimentation, became the prevailing force in the scientific world. While this was the case with sciences such as physics, the fields of medicine and biological science frequently had to rely on observation alone, meaning that many misconceptions from previous times were more difficult to disprove initially than in the other sciences.

One of the first seventeenth century studies of the lymphatic system was performed by the Italian physician Gaspare Aselli (1581–1625). Aselli was the first to formally describe the lymphatic vessels, or lacteals, of the intestines, although others had mentioned these structures previously. In his dissection of canine digestive systems, Aselli noticed that the lacteals in the intestine swelled after a meal, and that they contained a cloudy fluid (now known to be lymph). This study, conducted in 1622, was before William Harvey's influential work on blood circulation. As a supporter of Galen, Aselli failed to recognize the true role of the lacteals, and rather suggested that they were moving the food to the liver to manufacture blood. It is also interesting to note that Aselli's work was important because it included the first use of color graphics, and important advance in biological education.

Aselli's work was expanded on by the French physician Jean Pecquet (1622–1674). Pecquet was the first to discover that, in a dog, the lacteals drained into a central reservoir, called the *receptaculum* or *cisterna chyli*. He was also the first to discover that the thoracic duct, a main lymphatic vessel of the thoracic cavity (see Chapter 2), joined with the circulatory system at the superior **vena cava**. His work, pub-

Portrait of Dr. William Harvey, with an arterial chart in the background. © National Library of Medicine.

lished in 1651, was one of the leading investigations on lymphatic anatomy in the first part of the seventeenth century.

In the biological sciences, the start of the Scientific Revolution is frequently identified with the work of the English scientist William Harvey (1578–1657). Harvey's work on the circulation of the blood drastically challenged the ideas of human circulatory physiology that had persisted since the times of Galen. By 1628, Harvey had demonstrated that the blood was not manufactured by the body and consumed by the tissues, but rather circulated freely throughout the body. However, Harvey did not accept that the lymphatic system was a separate circulatory system, and instead incorrectly believed that the blood was the sole circulatory system in the body.

The first scientist to claim that the lymphatic system represented a separate circulatory system was the Danish physician Thomas Bartholin (1616–1680). In 1652, Bartholin published *De Lacteis Thoracsis*, in which he first uses the term lymphatic system. While Bartholin is given credit for naming the system, the most descriptive work on the physiology of the system in the late seventeenth century was conducted by the Swedish physician Olof Rudbeck (1630–1702). In 1653, Rudbeck published the results of his studies of the lymphatic system in hundreds of animals, including humans. His work supported Bartholin's claim that the lymphatic represented a second circulatory system. It was Rudbeck who discovered that lymph flows from the liver and thus is not carrying food directly from the intestines to the liver as had been previously thought. Later in the century, other scientists, such as Francis Glisson (ca. 1597–1677), provided evidence that the fluid of the lymphatic system may also be responsible for lubricating the major cavities of the body.

The microscopist Marcello Malpighi (1628–1694) suggested that the lymph nodes were part of the lymphatic system and not a form of fibrous growth as had originally been thought. He also recognized that the spleen is associated with the lymphatic system. In 1677, Johann Peyer (1653–1712) described a series of glands, now known as Peyer's patches (see Chapter 3), along the length of the intestines. Other scientists also made contributions, mostly in the description of the locations of lymphatic vessels and glands. By the end of the seventeenth century, there was developing in the scientific community the understanding that there were three major types of vessels in the human body: the arteries and veins of the circulatory system, and the vessels of the lymphatic system.

THE EIGHTEENTH CENTURY: THE ABSORBENT NATURE OF THE LYMPHATIC SYSTEM

Overall, the eighteenth century was a slow time in the study of the lymphatic system, although there were some important discoveries. Many references credit the discovery of the lymphatic system to two eighteenth-century physicians, William Hunter (1718–1783) and Alexander Monro (1733–1817). However, the work of Bartholin and Rudbeck in the seventeenth century had already demonstrated that the system existed. Monro and Hunter, working independently (ca. 1752–1754), were the first to discover that the lymphatic system has a role in the return of fluids from the tissues of the body. By injecting dyes into the circulatory system, both men also experimentally demonstrated that the arteries and lymphatic vessels were not directly connected, although there remains some controversy over who conducted the first of these experiments. Their works stated that the role of the interconnected vessels discovered by the seventeenth-century scientists was to drain fluids from the tissues and deposit the fluid in the circulatory system. This same idea had been suggested by Glisson in the seventeenth century, although he did not provide any direct evidence to support his claims.

One of the most accurate illustrations of the lymphatic system was made in 1787 by Paolo Mascagni (1752–1815). His work contained over forty illustrations of the lymphatic system, and was considered one of the most influential reference sources on the system well into the nineteenth century.

Some of the easiest cells in the body to identify are the white blood cells, or leukocytes (see Chapter 1). Although the use of the term leukocyte did not begin until 1855, a number of researchers reported their presence in the eighteenth century. One of these was William Hewson. Hewson correctly observed that white blood cells were formed by the lymphatic system in the lymph glands, and then transferred to the circulatory system through the thoracic duct. However, he incorrectly thought that some white blood cells were generated from red blood cells by the action of the spleen.

Another area of research that began in Western medicine during the late eighteenth century was the development of vaccinations. Credit for the development of vaccines is frequently given to the English physician Edward Jenner (1749–1823), although, non-Western cultures had also been experimenting with forms of vaccinations in ancient times (see earlier section "The Ancients").

Jenner was aware that there was a similarity between a disease called *cowpox* and the disease smallpox. However, unlike smallpox, which has a mortality rate of 30 percent, cowpox is rarely fatal. Furthermore, individuals who contracted cowpox appeared to be immune to the effects of smallpox. To validate this observation, in 1796 Jenner intentionally infected an 8-year-old boy with cowpox, and then two months later injected him with smallpox; the child did not develop smallpox. Jenner developed the term *vaccination*, from the Latin word for cowpox, to describe the process. The process quickly gained

Paolo Mascagni. © National Library of Medicine.

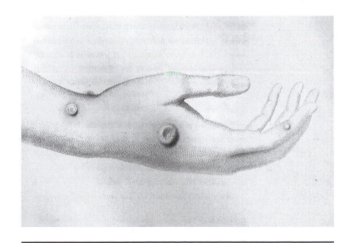

Hand of the Gloucestershire milkmaid showing the cowpox blisters from which Edward Jenner developed his smallpox vaccination technique. Courtesy of the Library of Congress.

acceptance among doctors in England, and over the next several decades is credited with greatly reducing the spread of smallpox.

THE NINETEENTH CENTURY: WHITE BLOOD CELLS AND IMMUNIZATIONS

The development of microscopic techniques and equipment in the seventeenth century by scientists such as Marcello Malpighi, Anton van Leeuwenhoek (1632–1723), and Jan Swammerdam (1637–1680), and the improvement of optics in the eighteenth century, provided the means for later scientists to examine the cells of the body. Many of the discoveries associated with the lymphatic system in the early eighteenth century focused on identifying and classifying the cells of the immune system.

A significant amount of attention in the nineteenth century focused on the white blood cells, or leukocytes. In 1843, the English physician Thomas Addison (1798–1866) described the movement of white blood cells out of the capillaries and into the surrounding tissues as part of the inflammation response (see Chapter 4). He noticed that the white blood cells had the ability to squeeze through the incredibly small gaps between the cells of the capillaries, a process that he called *diapedesis*. The amoeboid movement of these cells was described in greater detail by a number of other scientists in the mid-nineteenth century.

White blood cells destroy bacteria by the process of phagocytosis (see Chapters 1 and 4), in which projections of the cell membrane encircle the pathogen and move it into the cell, where the pathogen is then destroyed by digestive enzymes. The first accurate description of this process in white blood cells was provided by the Russian scientist Elie Metchnikoff (1845–1916). In his study of lower animals, Metchnikoff noticed that there existed special cells that ingested microscopic particles in response to tissue damage. When he turned his attention to the immune system of higher animals (ca. 1883), including humans, he noticed that white blood cells possessed a similar ability. He named these types of cells *phagocytes*, which is

METCHNIKOFF

Elie Metchnikoff. © National Library of Medicine.

Greek for "eating cells," the process by which the cells engulf the particles is called *phagocytosis*. Metchnikoff also played a significant role in the understanding of the immune response in the early twentieth century (see Chapter 7), and was awarded the Nobel Prize in physiology or medicine in 1908.

The study of the lymphatic system in the nineteenth century did not solely focus on cells. Another area of research was in the development of vaccinations. By the early nineteenth century, the use of cowpox to vaccinate against smallpox was widely used in England, but it was not until the development of the germ theory of disease by the French chemist Louis Pasteur (1822–1895) in 1862 that the relationship between microorganisms and disease was firmly established, and a dedicated effort could be made in protecting the population against the effects of certain diseases.

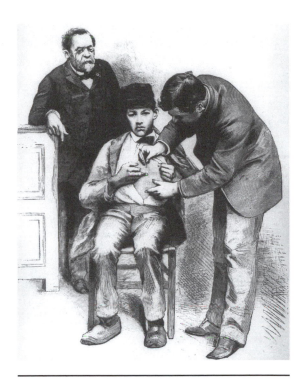

Louis Pasteur observes as a young boy receives an inoculation for hydrophobia. From *Harper's Weekly*, December 19, 1885. © National Library of Medicine.

For the remainder of the century, Pasteur remained one of the leading pioneers in the development of vaccines. Following his earlier successes with the germ theory of diseases, Pasteur was intensely interested in isolating the agents responsible for other common diseases. One of these diseases was cholera, a disease responsible for many forms of **dysentery**. As is frequently the case in science, his discovery was the result of a mistake in his scientific procedure. Pasteur was working with a highly lethal form of cholera in chickens, and on one occasion he mistakenly injected a group of chickens with a strain of cholera that had been prepared some time in the past. Whereas the cholera strain he was working with was normally lethal, this time the chickens displayed only minor symptoms of the disease and did not die. Furthermore, when he subsequently injected the same chickens with the potent strain of the disease, they showed no signs of contracting cholera. What Pasteur had discovered was the concept of attenuation, where the potency of a disease-causing agent is reduced so that it will produce an immune response, but not sickness. Pasteur used the same approach in his studies of the disease anthrax. Anthrax was a major problem of the farm industry, where it infected both farm animals and their human caretakers. In 1876, Robert Koch (1843–1910), a major contributor to the germ theory of disease, had isolated the bacteria responsible for anthrax, and by 1881, Pasteur had developed an attenuated strain of the bacteria that could be used to vaccinate both animals and humans against the disease.

To this point, Pasteur's work had involved developing vaccinations for diseases that had microbial causes. However, around 1885 Pasteur began work on a disease called *rabies*, which at the time was also called *hydrophobia*. Rabies was a challenge for Pasteur, because he was unable to isolate the bacteria responsible for causing the disease. It is now recognized that rabies is the result of a virus, but it was not until 1892 that Dmitri Ivanovski (1864–1920), and later Martinus Beijerinck (1851–1931), started accumulating evidence that these infectious agents existed. Because Pasteur was unable to isolate the bacteria for cultivation, he used rabbit hosts to generate material to produce his vaccinations. Once again he attenuated the potency of the vaccination by aging the infected material before injecting it into the patient. In 1885, Pasteur was successful in using the vaccine to prevent a case of rabies in a small boy suspected of being bit by a rabid dog.

As the nineteenth century drew to a close, there were several additional important advances in the development of vaccines. One of the prime problems with Pasteur's method was that it was exceedingly difficult to assess the potency of the attenuated vaccine without injecting it into a test subject. Because individual immune responses vary, there was always the chance that the vaccination may itself cause the disease. In 1890, a partial solution to this problem was provided by Emil von Behring (1854–1917). Instead of injecting attenuated material directly into the patient to provide immunity, von Behring first injected the material into an animal, and then isolated the animal's defensive compounds. These compounds, called *antitoxins*, could then be injected safely into the patient without risking infecting the patient with the disease. He is credited with using this procedure to develop vaccines against both **tetanus** and diphtheria.

Another important contributor in the development of vaccines was the chemist Paul Ehrlich (1854–1915). Ehrlich is considered to be the first immunochemist because of his intense interest in applying chemical principles to the study of the immune system. He made some of the most detailed descriptions of white blood cells in the nineteenth century. Ehrlich's contributions to the study of immunity span the turn of the century, and the majority of these are discussed in the next chapter. However, as an associate of von Behring, Ehrlich played a major role in determining the concentrations of antitoxin that were needed to provide immunity against a disease. The ability to deliver the correct dose of a chemical to fight a disease was an important step in the implementation of population-wide vaccination programs in the twentieth century.

7

The Lymphatic System in the Twentieth and Twenty-First Centuries

By the end of the nineteenth century researchers were beginning to put together the puzzle of the immune response. White blood cells had been discovered and the process of phagocytosis identified. In addition, the process of vaccinating populations against infectious diseases was gaining popularity, resulting in significant decreases in mortality from some diseases. However, it was not until the twentieth century that scientists began to truly understand how the lymphatic system and immune response function in fighting disease.

The twentieth century was a time of tremendous scientific advance. It is estimated that by the end of the century the amount of scientific information was doubling every five years. There can be little doubt that this pace will be maintained, if not accelerated, in the twenty-first century. The advances in the sciences of genetics and cell biology, the development of new procedures in medical technology, the formation of a new area called *computer science*, and the growth of pharmaceuticals, have all played an important role in the understanding of human physiology, including the function of the lymphatic system.

Gregor Mendel is often credited for founding the science of genetics, although in reality he was merely one of the first to formalize the mathematical basis of inheritance. Mendel's work remained relatively unknown until 1900, when three investigators—Erich von Tschermak (1871–1962), Carl Correns (1864–1933), and Hugo de Vries (1848–1935)—independently re-

A poster from the Chicago Department of Health advising inoculation against diphtheria, 1930s. Courtesy of the Library of Congress.

discovered the mathematical principles of inheritance that Mendel had established previously. While Mendel's work had very little influence on the science of the nineteenth century, the scientific climate had changed drastically by the start of the twentieth century. Charles Darwin's (1809–1882) theory of natural selection had sparked a tremendous interest in the natural sciences, specifically evolution, and many researchers were beginning to wonder how inheritance at the cellular and organism level worked. The rediscovery of the principles of transmission genetics set the stage for some of the most important discoveries in the history of mankind. Over the course of the century, genetics began to permeate all of the life sciences. Scientists began to predict the probability of a disease using statistics and **pedigree** analyses.

Diseases are typically the result of biochemical problems at the cellular level. Because the immune response is dependent on cells that travel throughout the system, biochemical problems are especially dangerous for the lymphatic system. Over the last thirty years, the advances in cell biology and genetics have revolutionized medicine. In cell biology, discoveries of cell signaling and development have unveiled some of the complex chemical signals that occur between the cells of the body, including the complex use of interferons and interleukins by the lymphatic system (see Chapters 1 and 4). In genetics, the revealing of the structure of DNA in 1953 by James Watson (b. 1928) and Francis Crick (b. 1916), the development of molecular biology in the 1980s and 1990s, and the relatively recent development of techniques to manipulate DNA in the laboratory, have all had a tremendous influence on the biological sciences and the study of medicine. One of the greatest achievements of science, the Human Genome Project, was completed in 2003. This advance will eventually enable the isolation of the genetic mechanisms of any disease, including those of the lymphatic system.

Advances in technology, specifically those to identify and treat diseases, also developed rapidly over the course of the twentieth century. At the end of the nineteenth century, physicians began to develop mechanisms to prevent disease, mostly through the development of vaccines and a greater understanding of the germ theory of disease (see Chapter 6). The use of radioactive **isotopes** to trace metabolic pathways in the body was made possible by late-nineteenth- and early-twentieth-century advances in the science of **nuclear medicine**. Coupled with this was the new ability to peer inside the human body using procedures such as **ultrasounds, magnetic resonance imaging (MRI)**, and **computerized axial tomography (CAT)** scans.

A new area of research called computer science originated in the twentieth century. Starting from a simple computer with limited abilities to conduct mathematical calculations in the 1940s, computers had developed into their now-recognizable form by the

Hugo de Vries, 1933. © National Library of Medicine.

1970s. In the last three decades of the twentieth century, computers began to have a tremendous influence on the life sciences. The invention of supercomputers with incredible capacities for data processing has enabled scientists to quickly perform calculations and analyses that previously were either impossible or prohibitively time-consuming. The expansion of the Internet from a military communication device to its present state as an information portal has placed invaluable resources at the fingertips of scientists. Online databases, document sharing, and practically instantaneous communication have all played an important role in the rapid scientific advances in medicine, physiology, and genetics in the past thirty years.

While the study of inheritance enabled researchers to determine patterns and mechanisms, and technology provided the diagnostic tool, pharmaceutical drugs have given physicians the ability to treat diseases of the lymphatic system. Chemotherapy, or the use of chemicals to destroy cancer cells in the body, began in the early twentieth century and remains a powerful mechanism for fighting cancer to this day.

It is beyond the scope of this work to provide a detailed examination of every advance in the study of the lymphatic system and immune response in the twentieth and twenty-first centuries. Instead, this chapter will focus on the major advances in the study of this system by following the Nobel Prize awards in the last century that relate directly to the lymphatic or immune systems.

NOBEL PRIZES

One of the greatest recognitions that may be bestowed upon a scientist is the Nobel Prize. Awarded to "those who, during the preceding year, shall have conferred the greatest benefit to mankind," the Nobel Prizes are given on behalf of the estate of Alfred Nobel (1833–1896). Nobel was a nineteenth-century inventor and chemist who was responsible for stabilizing the explosive compound nitroglycerine by combining it with diatomite; the result was dynamite. Nobel became rich from his invention, but despaired at being called the "merchant of death." The Nobel Prizes, awarded annually since 1901, with the exception of during some periods of international conflict, are Alfred Nobel's attempt to leave something beneficial for mankind.

In modern times, Nobel Prizes that are associated with the lymphatic system may be awarded either in the category of Physiology or Medicine, or in Chemistry. The Nobel Prizes that relate in some manner to the lymphatic system or immune function are listed in Table 7.1. A few of the most relevant of these are detailed in the following paragraphs.

THE DEVELOPMENT OF IMMUNOCHEMISTRY

At the start of the twentieth century, there were two major theories on how the body obtained immunity to infectious agents. The first is called the *humoral theory of immunity*. This theory focused on the belief that the blood, or some component in it, gave the body the ability to resist pathogens, how this happened was still unclear at the start of the twentieth century. Some believed that the blood sequestered infectious agents away from the tissues, and others believed that blood clots represented destroyed bacteria. However, after the discovery of antitoxins by Emil von Behring, and later work by Paul Erhlich (see Chapter 6), the majority of scientists considered that some component of the blood physically destroyed the invading organisms and their toxins, preventing the onset of disease.

Opposing the humoral theory of immunity was a second theory, called the cellular theory of immunity. This theory focused on the idea that the cells of the body were responsible for the destruction of the incoming pathogens. Elie Metchnikoff's nineteenth-century discovery of phagocytic cells (see Chapter 6) provided much of the evidence to support this theory. In his studies of the inflammatory response, Metchnikoff noted the phago-

TABLE 7.1. Nobel Prizes Relating to the Lymphatic System

Year	Recipient	Summary
1901	Emil von Behring	Development of antitoxins
1905	Robert Koch	Contributed to the development of the germ theory of disease
1908	Paul Ehrlich	Numerous contributions to the study of immunity, including the concept of chemotherapy
1908	Elie Metchnikoff	The role of phagocytes in the immune response
1913	Charles Richet	Allergic responses
1919	Jules Bordet	Action of antibodies and complement proteins in the immune response
1927	Julius Wagner-Jauregg	The use of inoculations to treat nervous system disorders
1930	Karl Landsteiner	Discovery of human blood groups and A/B antigens
1939	Gerhard Domagk	Development of antibacterial sulfonamide drugs
1945	Alexander Fleming Ernst Chain Howard Florey	Discovery of the antibiotic penicillin
1952	Selman Waksman	Discovery of the antibiotic streptomycin
1960	Macfarlane Burnet Peter Medawar	Immunological tolerance and the relationship to transplants
1972	Gerald Edelman Rodney Porter	Chemical structure of antibodies
1976	Baruch Blumberg	Screening of blood for hepatitis antigens and development of hepatitis vaccine
1980	Jean Dausset George Snell Baruj Benacerraf	Relationships between cell surface proteins, the immune response, and transplant effectiveness

TABLE 7.1. (*continued*)

Year	Recipient	Summary
1982	Sune Bergstrom Bengt Samuelsson John Vane	Work on prostaglandins
1984	Niels Jerne Georges Kohler Cesar Milstein	The specific nature of the immune system and the production of monoclonal antibodies.
1987	Susumu Tonegawa	Genetics of antibody diversity
1988	James Black Gertrude Elion George Hitchings	Development of several important drug treatments, including histamine blockers, antimalarial drugs, and antiviral drugs.
1990	Joseph Murray	The use of immunosuppressive drugs for transplantation
1996	Peter Doherty Rolf Zinkernagel	Cell mediated immunity

This table contains only those awards that relate directly to the function of the lymphatic system and immune response. Many other awards represent achievements that indirectly contributed to an understanding of this system.

cytic action of white blood cells and the destruction of bacteria. Although the early-twentieth-century scientists did not yet know it, both theories were correct. It is the combined nature of the humoral and cellular responses that gives strength to the action of the human immune system (see Chapter 4).

In the early twentieth century, there was a significant amount of attention being given to the function of the antitoxins. To the nineteenth-century scientists, an antitoxin was simply recognized as some chemical that counters the toxin being produced by the bacteria. In the late nineteenth century, one of the leading investigators of antitoxin function was Paul Ehrlich, a chemist who played an important role in the development of the science of **immunochemistry**. Ehrlich's first major accomplishment in the study of immunology occurred at the end of the nineteenth century, when he was the first to accurately measure and standardize the levels of toxins and antitoxins. This is widely recognized as being one of the first attempts to quantify some aspect of the immune system, and is considered to be an important advance in the development of correct dosages for both vaccinations and antitoxins.

While an antitoxin is only effective against a toxin for a relatively short period of time, the body does possess a longer lasting mechanism of protection against specific invaders. These are the antibodies, specialized proteins that neutralize pathogenic organisms and viruses either by destroying them directly, or targeting them for destruction by other cells of the immune system (see Chapters 1 and 4). By the end of the nineteenth century, scientists widely recognized that antibodies existed, although they had yet to determine that they were proteins or their specific mode of action. In the early twentieth century, the terms antitoxin and antibody were frequently used interchangeably.

Ehrlich is most widely recognized for his work on what is called the "side-chain" theory of antibody function. Ehrlich believed that the body contained a limited number of antibodies. These antibodies were specific in what they interacted with at the molecular level. Furthermore, once the antibody interacted with its target antigen, the bond was irreversible. Ehrlich also stated that antibodies were a product of

Paul Ehrlich. © National Library of Medicine.

the cell and that they were associated with the membrane of the cell. Initially Ehrlich believed that the antibody was a receptor on the membrane of the cell, although he later modified this theory to account for conflicting observations on how some aspects of the immune system worked. To explain his side chain theory, Ehrlich used a graphic representation of the cell membrane, with the receptors interacting geometrically with antigens (see Figure 7.1). Although considered by many to be a crude portrayal of the immune response, it is also considered to be one of the most influential diagrams of a cellular process in the early twentieth century. Even without a scientific background, it is possible to understand the specificity of the antibody to the antigen based on Ehrlich's diagram. For his outstanding contributions to the study of the immune system, Ehrlich shared the 1908 Nobel Prize in physiology or medicine for his work. The other recipient was Elie Metchnikoff, the discoverer of the phagocytic action of white blood cells.

Another researcher who was investigating antibody function in the early twentieth century was the Belgian scientist Jules Bordet (1870–1961). Bordet did not accept Ehrlich's initial interpretation of antibody function, but rather

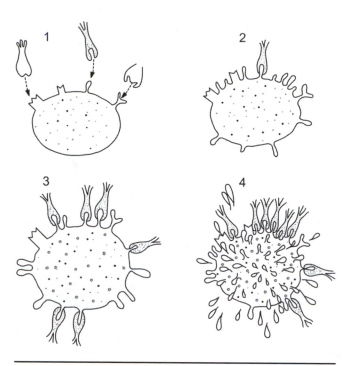

Figure 7.1. Ehrlich's "side-chain" theory of antibody function.

This reproduction of Ehrlich's original drawing depicts the interaction of antibodies with specific part of the cell.

believed that antibodies rid the body of antigens by "complementing" with them. Complementation was believed to be a physical combination of the antibody (or antitoxin, as it was called in the early twentieth century) and antigen, which prevented the antigen from causing the associated disease.

Bordet's ideas in complementation were derived from his experiments with bacteriolysis. The word *lysis* means "to break," and thus bacteriolysis is the breakdown of bacteria by the immune system. Experiments indicated that if a rabbit's blood is injected into a guinea pig, the rabbit blood is destroyed by some compound in the guinea pig. The **serum** from the guinea pig was found to be toxic, if not lethal, to the rabbit. This process is called *hemolysis*, or "blood-breaking." Bordet discovered that if the guinea pig serum was heated, usually at temperatures over 55°C, then the serum lost its toxic characteristics. After interpreting his data, he believed that there was a two-step process at work. First, the blood cells became sensitized by the immune system. The sensitized cells were then targeted for destruction by the complement system. Bordet believed that the complement system was a general (or nonspecific) system, but others thought that the complementation reactions demonstrated specificity.

In the early twentieth century, Bordet expanded these principles to the study of the immune response against bacteria, believing that the same general process is responsible for the destruction of bacteria within the body by the immune system. Initially, he thought that these systems were the same, but some of his later experiments demonstrated that there were actually two systems at work, a hemolytic and bacteriolytic system. Futhermore, through experiments with the plague, anthrax, and typhoid fever, Bordet and his colleagues discovered that it was possible to use the differences between these systems as a diagnostic tool to determine if an organism had been previously exposed to the pathogenic organism responsible for the disease. For his work on the complementation system, Bordet received the 1919 Nobel Prize in physiology or medicine.

In 1906, August von Wassermann (1866–1925) and Albert Neisser (1855–1916) used Bordet's work to successfully develop a test to determine if a person had been exposed to the bacteria responsible for **syphilis**. This test, called the *Wassermann Reaction* (or *Wassermann Test*) was one of the first diagnostic tools of its kind, and over the next several decades several researchers modified the technique to make it both more sensitive and easier to perform.

Bordet's complement system was not the only diagnostic procedure developed from the study of immunochemistry in the early twentieth century. From some of Metchnikoff's work in the late nineteenth century, it was known that chemicals in the immune system caused bacteria to clump together. This process is called *agglutination*, and the substances that cause the reaction are called *agglutinates*. Herbert Durham (1866–1945) and Max Gruber (1853–1927) are recognized as being the first to characterize these reactions, which quickly led to the development of a diagnostic procedure using agglutination for the presence of the bacteria responsible for typhoid fever. A second type of reaction is the precipitation reaction, in which a **precipitate** forms upon exposure to a pathogenic organism. Because these precipitates are frequently very specific in their chemical nature, they can be used to detect the presence of the organism in the blood sample.

The early twentieth century was a time of tremendous discovery in the field of immunochemistry. During this time, many of the fundamental chemical processes of the immune response were actively investigated. Research in this important area of immune function continues to this day, and some of the more important discoveries are contained in the following sections.

ANTIBODIES

One important aspect of immunochemistry that deserves special attention is the study of antibodies. Antibodies are proteins produced by B cells (see

Chapters 1 and 4) as part of the specific defense mechanisms in the body. The work of Ehrlich and Bordet, as well as the process of agglutination and precipitation, are largely due to the action of these antibodies. The fact that antibodies were proteins was widely agreed upon by the scientific community, because it was believed that, among the biologically important molecules, only the proteins had the structural complexity to be able to interact with the antigens.

One of the leading researchers on antibodies in the first half of the twentieth century was Karl Landsteiner (1868–1943). Landsteiner discovered human blood groups (1900) and Rh factors in the blood (1940). For these achievements, he was awarded the Nobel Prize in physiology or medicine in 1930. Although these antigens are important in the function of the circulatory system and the process of human transfusions (see the Circulatory System volume for more information), Landsteiner himself thought that his most important contributions to science was his study of antibodies.

Landsteiner was one of the first to experimentally demonstrate the specific nature of antibodies. Landsteiner used a group of chemicals called *haptens* for his study. A hapten is a molecule that is combined with a protein to invoke an immune response. These are not true antigens because the molecule and protein are typically not associated with a disease. For example, Landsteiner combined arsanilic acid with egg albumin. When injected into an animal, the animal's antibodies would interact with the hapten, but not with the individual components. Landsteiner suggested that it may be possible to generate vaccines by combining a pathogenic microorganism with a protein in such a way that antibody production would be initiated, but the microorganism would lose its ability to cause a disease. It would be almost half a century before this process would be technically possible.

Landsteiner had demonstrated that antibodies were specific, but that did not explain exactly how they functioned. Ehrlich had previously thought that the number of these molecules was very limited. This presented some problems, because there was no conceivable mechanism for a body to predict all of the possible antigens that it would be exposed to during its lifetime. Some, such as Linus Pauling (1901–1994), believed that the body contained some generalist proteins, which, when exposed to an antigen, formed a shape specific to the organism being targeted.

A portion of this debate was resolved in the early 1970s by Gerald Edelman (b. 1929) and Rodney Porter (1917–1985) who independently determined the linear sequence of the amino acids in an antibody. For their work they shared the 1972 Nobel Prize in physiology or medicine. Their work enabled other researchers to identify the hypervariable regions of the antibody (see Chapter 1), further demonstrating that there is tremendous diversity in antibody structure.

By this time, scientists recognized that proteins were produced by specific instructions in the genetic material. So the specificity of antibodies produced a challenge for geneticists. Because an organism, such as a human, did not possess an endless source of genes, how could the almost limitless potential for antibody diversity be explained? The answer was provided in 1976 by the Japanese geneticist Susumu Tonegawa (b. 1939). Tonegawa demonstrated that the cells of the immune system, the B cells specifically, had the ability to shuffle the genetic information responsible for antibody formation into an almost endless number of combinations. For this work, Tonegawa received the 1987 Nobel Prize in physiology or medicine.

Research investigations on a specific disease are complicated by the fact that the body may generate many different forms of antibodies against the same pathogen. This problem was solved by the invention of a system of producing identical antibodies, called *monoclonal antibodies*, in 1975 by Cesar Milstein (1926–2002) and Georges Kohler (1946–1995). For their discovery, which has had a tremendous influence on immunological research and the study of proteins, they shared the 1984 Nobel Prize in physiology or medicine. A third researcher, Niels Jerne (1911–1994), also received a portion of the Nobel Prize that year for his development of theories of how antibodies work in the context of the immune response.

ANTIBIOTICS

An antibiotic is any substance produced by an organism that either inhibits the growth of microorganisms or destroys them completely. The term is actually derived from the concept of antibiosis, first described scientifically by Paul Vuillemin (1861–1932) in the nineteenth century. Antibiosis is when one creature kills another to preserve its own existence. The term *antibiotic* is believed to have first been used by Selman Waksman (1888–1973). While many may consider antibiotics a product of modern medicine, in reality the antibiotic properties of many chemicals have been known for millennia. The ancient Chinese, as well as medical sources from the Greeks and Romans, indicated that many molds (fungi) have antibiotic properties, but in many cases these were simply regarded as examples of folk medicine. It is interesting to note that many of our modern antibiotics are derived from the fungi, which naturally are highly resistant to microbial infections.

Many history texts indicate that Pasteur recognized that there were antimicrobial properties present in natural compounds, although he did not specifically use these to combat a disease. The modern age of antibiotic discovery is said to have started around the time of World War I with the work of the Scottish bacteriologist Alexander Fleming (1888–1955). Prior to Fleming's work, a number of researchers had noted that under the right condi-

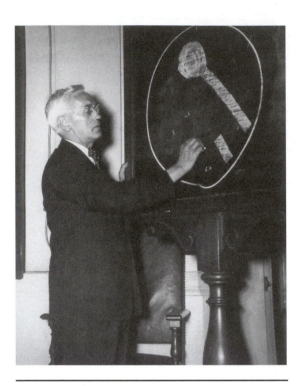

Alexander Fleming. © National Library of Medicine.

tions colonies of microbes tended to disappear, which suggested that something in their environment was causing them to lyse. One of Fleming's first advances was the discovery of lysozyme, a chemical naturally produced by the body that destroys bacteria. Lysozyme is naturally produced by a number of the tissues of the body, including the skin, and although it is not technically an antibiotic, it did lead Fleming to think that similar compounds may be found in other organisms.

In 1928, Fleming was working with a pathogenic strain of staphylococci bacteria when he discovered that a certain mold present in the lab was causing bacterial colonies to disappear. Interestingly, he noted that the fungi did not appear to harm human tissues, only bacterial colonies. The organism was *Penicillium notatum*, and the compound it produced to destroy the bacteria is the antibiotic penicillin.

Fleming did not capitalize on the potential of this antibiotic, but in the late 1930s a research team led by Howard Florey (1898–1968) and Ernst Chain (1906–1979) began to work on the isolation of penicillin. Initially there were considerable difficulties producing penicillin, because the fungus only creates a limited amount of the antibiotic. Furthermore, because this was the first antibiotic to be recognized, mass production facilities did not exist. After concluding that the antibiotic would treat human infections, the researchers approached several pharmaceutical companies and by the early 1940s the drug was being mass produced. The timing could not have been better; World War II was underway and war injuries were plentiful. The availability of penicillin is recognized as a major factor in the survival of many injured military personnel during the course of the conflict and has saved countless lives across the world since its discovery. For their work on penicillin, Fleming, Florey, and Chain shared the 1945 Nobel Prize in physiology or medicine.

Another important class of antibiotics is streptomycin. Credit for their discovery is given to Selman Waksman. Initially, Waksman worked with a chemical from fungi in the soil called *actinomycin*. This chemical proved to be toxic to humans (and other mammals), but in 1944 he discovered the streptomycins. Also derived from a fungus (*Streptomyces griseus*), this antibiotic played an important role in limiting the severity of tuberculosis. Waksman received the 1952 Nobel Prize in physiology or medicine for his discovery.

Since these early discoveries, a number of antibiotics have either been discovered or synthesized in the lab. However, one of the greatest challenges to modern medicine is the resistance of microbes to antibiotics. Antibiotic resistance is the result of metabolic or physiological changes on behalf of the target microorganism in response to exposure. Evolutionary biologists consider the process by which antibiotic resistance forms to be a model of evolution by natural selection. Microbes in a population have some degree of genetic and metabolic diversity. The use of an antibiotic may kill susceptible microbes, but those with altered metabolism or physiology may survive, passing their traits on to the next generation. Because bacteria may divide very quickly, within a short period of time all of the microbes in a population may be resistant to the initial antibiotic.

The widespread use of antibiotics, and the misuse of them by the general public and agricultural industries, has reduced the effectiveness of many antibiotics. Scientists are now actively looking for new classes of antibiotics in fungi, insects, and plants. Efforts are also underway to reduce the use of antibiotics in agriculture, in an attempt to prolong the life of the existing chemicals.

ORGAN TRANSPLANTS AND THE IMMUNE RESPONSE

One area of scientific research that received a significant amount of attention during the second half of the twentieth century was the study of how the immune system responds to organ transplants. Since the 1950s, medical professionals have had the ability to transplant organs from a donor to a recipient, but the problem of tissue rejection reduced the effectiveness unless the donor and recipient were closely related.

One of the first individuals to study the response of the immune system to a transplant was the Australian bacteriologist Macfarlane Burnet. Burnet suggested that it was possible to take embryonic tissue and expose it to foreign tissue. The embryonic tissue would then effectively be immunized against the effects of transplantation and then could be placed into a donor. The first successful use of this procedure was done by the English scientist Peter Medawar, who experimented with transplantation using the embryos of mice. These individuals shared the 1960 Nobel Prize in physiology or medicine for their work.

The major histocompatability complex (MHC) markers on the surface of the cell membrane provide the mechanism by which the body identifies tissues that belong to "self." Early study on these receptors was conducted by

George Snell (1903–1996), who demonstrated that the MHC markers were actually a **polygenic** system. This work was expanded upon by both Jean Dausset (b. 1916) and Baruj Benacerraf (b. 1920), who independently contributed to the understanding of the genetic structure of MHC markers. This information paved the way for the development of immunosuppressive drugs and the subsequent increase in transplant success rate. These individuals shared the 1980 Nobel Prize in physiology or medicine for their efforts.

MODERN ADVANCES

In the past few decades the pace of scientific advance with regards to the function of the immune response has accelerated rapidly, especially within the past ten years. The science of immunology represents an area of intense research, with organizations such as the National Institutes of Health (NIH) actively funding projects to increase our understanding of human disease. By some estimates the amount of information on the immune system is doubling every three to five years, an astonishing figure considering where science has come from since the start of the twentieth century. Almost on a quarterly basis, top scientific journals such as *Nature* and *Science*, report findings related to the immune system. For those who wish to keep up with these advances, but lack the technical background for understanding major scientific journals, public-oriented publications such as *NewScientist* and *Scientific American* frequently carry news information and reviews of major advances in the science of immunology.

POSSIBILITIES FOR THE FUTURE

It is difficult to predict the future of scientific advance because the development of a single new experimental procedure can alter the path of scientific study almost overnight. It is possible, however, to examine some of the trends in scientific research and extrapolate as to where current studies may take us.

The science of molecular biology has significantly changed the nature of science. A little over fifty years ago, the structure of the double helix was first determined. Since that time, scientists have developed the means to manipulate genetic systems in the lab. In the early 1990s, the idea of mammalian cloning and mapping of the entire human genome still bordered on science fiction, and yet today these goals have been achieved. Scientists now have at their disposal massive databases of genetic and biochemical information that significantly enhance their ability to study the relationship between genetic defects and a disease state. While historically it used to take decades to link a disease to its cause, the process can usually now be accomplished in a few years. This greatly accelerates the development of treatments or cures.

Currently there is an intense interest in developing the capability to use the immune response effectively against malfunctioning cells of our own body, such as cancer cells. Because cancer cells represent "self," there have been obstacles in targeting immune cells against them, although it is known that some classes of immune cells, such as cytotoxic T cells (see Chapter 1), may naturally destroy tumor-causing cells. Many researchers also believe that it may soon be possible to genetically engineer immune system cells outside of the body and then introduce them into the system to destroy potentially fatal illnesses, including cancer. Other scientists are actively working to improve the signaling systems of the human immune system, such as interferons (Chapter 1), to combat viral infections. Antiviral medications, still in their infancy, now delay the ability of viruses to replicate, giving the immune system time to destroy the viruses. The next decade should see an increase in their specificity and ability to fight disease.

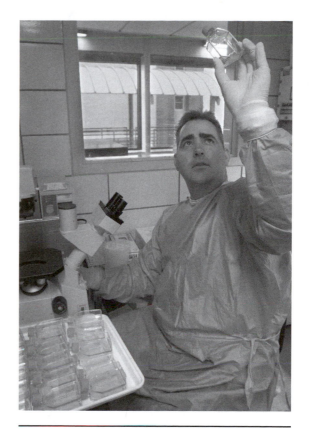

A researcher at the Special Pathogens Branch of the CDC examines a T-25 flask used in SARS virus isolation, 2003. © Centers for Disease Control and Prevention.

Despite these advances, the science of immunology will not become obsolete in the near future. New pathogens regularly appear on the world scene and, in the process, challenge researchers to rapidly find a cure, preventative mechanism, or treatment. One such example is the sudden acute respiratory virus (SARS), which although it did not kill as many people as influenza or AIDS, paralyzed the international community in the spring of 2003. Such outbreaks of new diseases are bound to become more common as humans interact on a more crowded earth.

8

Diseases of the Lymphatic System: Cancer

The next two chapters will focus on diseases and disorders that affect the various parts of the lymphatic system. While Chapter 9 will feature a brief overview of these disorders as they relate to primary and secondary organs, this chapter will focus on cancer. A disease that affects all body systems, cancer is especially dangerous to the lymphatic system, which protects the body against disease and infection. Many types of cancers even start in the lymphatic system. While the lymphatic system is designed to fight infection, it is powerless to fight cancer on its own. However, testing and research have produced many effective treatments that enable men and women to combat cancer and go on to lead healthy lives.

The second leading cause of death in the United States, cancer will affect nearly 50 percent of men and one-third of all women some time during their lifetimes. At a basic level, cancer develops when cells located in a particular part of the body begin to grow out of control. While there are many different kinds of cancer, which will be discussed in this chapter, each and every type is characterized by this abnormal cell growth.

During a person's growth years, which begin at birth and can extend to early adulthood, cells in the body grow, divide, and then die at an orderly pace. This pattern happens more rapidly in the earlier years of one's life, but cells in most parts of the body slow down in the early stages of adulthood. Once this pattern slows down, cells divide and thus reproduce only when replacing cells that have died or worn out, and when healing injuries. When a person has cancer, his or her cells grow and divide, but do not die, making them distinct from normal cells. These cancer cells outlive healthy, normal cells and continue to divide, forming even more abnormal cells.

Cancer cells develop due to irreparably damaged DNA. When DNA is damaged during normal development, the affected cells either die or are able to repair themselves. What makes cancer unique—and incurable—is that the damaged DNA cannot repair itself. In some cases, cancer patients have inherited this damaged DNA, or have harmed DNA through smoking (see "Smoking and Cancer") But there are many cases, where cancer can appear in someone who hasn't smoked and who doesn't have a family history of the disease.

Most types of cancers begin forming as **tumors**. Not all tumors are cancer-

Smoking and Cancer

In 1982, the U.S. Surgeon General issued a report declaring cigarette smoking to be the major cause of cancer-related deaths nationwide. Cigarette smoking causes about 87 percent of lung cancer deaths. In addition, smoking has also been linked to cancers all over the body, including those in the esophagus, mouth, bladder, and stomach, in addition to certain cancers of the lymphatic system, including leukemia. In addition to cancer, smoking is a leading cause of disorders such as heart disease, bronchitis, emphysema, stroke, and is a contributing factor to severe pneumonia. Smoking is especially harmful to women, who have an increased risk of miscarriage and infant death, along with low birth weight, if a woman smokes while she is pregnant. Studies have shown that cigarette use eventually kills half of all smokers, and cigarettes cause more deaths in America than AIDS, alcohol, suicide, car accidents, homicide, and illegal drugs. Cigarette smoking is an acquired behavior, making it the most preventable cause of premature death in the United States.

Tobacco is the primary ingredient in cigarettes and cigars. These tobacco products (as well as smokeless and pipe tobacco) contain more than 4,000 individual compounds, 43 of which are **carcinogens**. In addition, manufacturers also supplement tobacco with harmful substances such as ammonia, tar, and carbon monoxide in order to enhance the flavor.

Many smokers who claim to be addicted to cigarettes, cigars, or smokeless tobacco are addicted to the drug **nicotine**, which is a primary ingredient in tobacco.

Addiction is defined as the repeated or compulsive use of a substance, even though it may be harmful. Six years after smoking was declared the leading cause of cancer deaths in the country, the U.S. Surgeon General stated that nicotine is the drug in tobacco that causes addiction; cigarettes and other forms of tobacco (cigars, smokeless and pipe tobacco) are addicting; and the behavioral and drug-related properties that determine tobacco addiction are similar to those that determine addiction to drugs, such as heroin and cocaine. Studies have shown that the daily use of tobacco products causes addiction in a proportionate number of users.

ous—these are called **benign**—but tumors that are determined to be cancerous are called *malignant.* Benign tumors do not spread to other parts of the body, and in almost all instances, are not life threatening. There are a few types of cancer that do not form tumors, such as leukemia. Leukemia cancer cells form in the blood and blood-forming organs, and then grow while circulating through the body's tissue network. In all kinds of cancer, the abnormal cells spread throughout the body, where they grow and eventually replace the normal cell tissue—a process called **metastasis**, which begins when the cancer cells first enter the blood stream or lymph vessels. Cancer is characterized by where the cancer cells begin to form and grow. For instance, a patient might have breast cancer that has metastasized to the liver. It is still considered breast cancer, not liver cancer (see "Cancer Testing").

Because each type of cancer behaves differently, treatment is tailored to the specific diagnosis. The remainder of this chapter will focus on the various types of cancer associated with the lymphatic system, followed by the different kinds of tests and treatments for cancer (see "Staging"). However, it is first important to have a basic understanding of the two primary types of treatment used to help the body fight cancer cells: **chemotherapy** and **radiation therapy**. Chemotherapy is treatment through the use of medications that kill cancer cells. These drugs are ingested through the mouth in pill or

Cancer Testing

Some common tests for cancer include the CT scan, the MRI, and the Gallium and bone scans. These tests produce specialized images of the body in order for doctors to view the density and location of the cancer tumors.

Computerized tomography (CT): When patients undergo a CT screening, they lie on their back on a table, which moves slowly and continuously through the center of a doughnut-shaped scanner containing the x-ray beam. These x-rays scan the subject from all angles as they move through the scanner, while a computer generates what can ultimately be a three-dimensional model of the different regions of the body.

Magnetic resonance imaging (MRI): Using large magnets and radiowaves, this test uses a computer to produce images of internal organs. These computer-generated images look similar to images produced through the CT testing, but produce a more accurate image of lymph node enlargements.

Gallium scan and bone scan: This imaging technique begins with an injection of a radioactive chemical into a vein. This chemical is attracted to areas of the body where cancer is present, and an image of these areas can be viewed through a special camera, including in the bones and internal organs. The Gallium scan is best suited to finding tumors that are growing at an aggressive rate.

Staging

Staging is defined as the process by which doctors determine how much cancer is in a patient's body and the location of the cancer. There are two reasons why doctors classify cancer into stages: it is a way of gathering information to determine the best course of treatment, and so doctors can communicate the prognosis with patients.

There are four criteria a doctor will look at to determine a cancer's stage. One is the size of the tumor, while another is whether or not the tumor has grown into neighboring areas of the body. The third criteria is whether the cancer has metastasized or spread to surrounding lymph nodes, while the final determination is whether it has spread to distant areas of the body. In addition, doctors look at four types of evidence in order to classify the cancer at different sites and different periods of time.

- *Clinical-diagnostic staging:* The doctor will determine the amount of cancer through feeling and observing the tumor, in addition to using x-rays and other tests.

- *Surgical-evaluative staging:* This evaluation is made after exploratory surgery and/or a biopsy.

- *Post-surgical treatment pathologic staging:* The tumor is examined after surgery and its cancerous cells are analyzed using a microscope.

- *Re-treatment staging:* This determination is made after new or additional treatments are implemented.

After this aspect of staging is determined, the cancer is classified according to the TNM system, which was developed as a tool for doctors to determine the extent of the cancer based on standard criteria. The "T" stands for primary tumor, which is then assigned a number between 1 and 4 that describes the tumor's size and the extent to which it has spread to neighboring areas of the body. If there is no evidence of a tumor, a "0" is assigned. The second letter, "N," stands for lymph nodes, which are also assigned a number 0 to 2 that indicates whether the tumor has spread to the local lymph nodes, in addition to the size of the nodes and the number of nodes affected. Just like the tumor classification, a "0" means that the regional lymph nodes do not appear to be abnormal. The final classification letter, "M" evaluates the presence of metastases, or whether the cancer has spread to other regions of the body. This letter is accompanied by a 0 or 1; a 0 means that there are no detected distant metastases and 1 means that distant metastases are present.

Each of these classifications is combined, and a stage is assigned—I, II, III, or IV. Each tumor and cancer has its own classification system, so the letter and number combination can have different meanings depending on the cancer. In general, however, stage I cancers have the better prognosis for treatment and survival, because they are less advanced. The most advanced cancers are considered stage IV.

liquid form, or they can be injected into a vein located in a muscle or under the skin. Chemotherapy is considered systemic therapy because it circulates throughout the bloodstream in order to travel to the cancer cells. But while chemotherapy medications kill cancer cells, they also can harm normal cells. These drugs specifically target rapidly dividing cells, which is why they are effective in targeting cancer cells that divide and reproduce at a faster rate than normal cells. However, cells in the bone marrow, mouth, and intestinal lining, in addition to hair follicles, also grow at a rapid rate in order to replace cells that quickly wear out. Therefore, these cells are often affected by chemotherapy, causing hair loss, mouth sores, lowered immunity (due to low white blood cell counts), excessive bruising, and fatigue (due to low red blood cell counts). Appetite loss, nausea, and vomiting are common side effects of chemotherapy caused by damage to the intestinal cells, in addition to the effect of the medications on certain areas of the brain that control the appetite.

Through the use of high-energy rays, radiation therapy attempts to decrease the growth rate or kill cancer cells. The radiation beam is delivered from a machine outside the body, known as external beam radiation. This method is often used when the cancer is localized in one part of the body, and is sometimes used when the cancerous tumor region is unmanageable by chemotherapy alone. Radiation has been shown to be most successful when applied to only those areas in the body that are affected by the cancer. When used in combination with chemotherapy, the treatment is known as involved field radiation, which is now the preferred method of radiation therapy. The patient is given three to four courses of chemotherapy, and then radiation will be implemented to target those areas of the body where the amount of cancer is greater.

The external beam radiation has side effects similar to chemotherapy. Both therapies are similar in that the drugs and radiation beams can harm healthy tissue located near cancerous cells. Fatigue, nausea, and diarrhea are three common side effects.

LEUKEMIA

Leukemia is a type of cancer that begins in the body's supply of blood cells. Under normal circumstances, blood cells form in the bone marrow, which is the soft material found in the center of most bones. Stem cells and blasts are immature blood cells, which then mature in the blood marrow and eventually move into the blood vessels. As explained in previous chapters, the bone marrow makes three different types of blood cells. The white blood cells help the lymphatic system fight infection, the red blood cells transport oxygen throughout the body's tissue network, and the **platelets** help form **clots** that control blood flow. When people have leukemia, their bone marrow makes abnormal white blood cells, which are referred to as leukemia cells. While these cells might function normally at first, they soon begin to dominate and crowd out healthy white blood cells, red blood cells, and platelets, making it harder for the blood and circulatory system to function properly.

There are two classifications for leukemia: chronic and acute. When a patient has chronic leukemia, his abnormal blood cells may not have any symptoms during the early stages of the disease. In fact, the abnormal blood cells may still perform their function. But over time, chronic leukemia gets worse as the number of infected cells in the blood rises. Acute leukemia is characterized by very abnormal blood cells that are never able to function properly. As the number of leukemia blood cells increases at a rapid pace, the disease quickly worsens.

Leukemia is also classified among four types, based on the type of white blood cell that is affected. This type of cancer can develop in lymphoid or myeloid cells. Chronic lymphocytic leukemia most often occurs in people over the age of 55, and chronic myeloid leukemia also affects mainly adults. Acute lymphocytic leukemia is the most common type of the disease found in children, while acute myeloid leukemia can occur in both adults and children.

Another type of chronic leukemia is called *hairy cell*. This type of leukemia affects the white blood cells that are known as *lymphocytes* (see Chapter 1). Hairy cell leukemia is characterized by short, thin projections that look like hairs that grow on the surface of the lymphocytes. These cells gather in the bone marrow, spleen, and lymph nodes, and crowd out healthy white blood cells, particularly in the bone marrow.

Symptoms and Diagnosis

Some common symptoms of leukemia can include swollen lymph nodes (especially in the neck or armpit areas), fevers or night sweats, frequent infections or illness, fatigue or exhaustion, headache, extensive bleeding and bruising, and pain in the joints or abdomen. Other symptoms can include vomiting, loss of muscle control, and seizures. In male patients, leukemia cells can also build up in the testicles, which can lead to swelling. Some patients also develop sores in the eyes or on the skin. It is important to note that these symptoms are not definitive signs of leukemia, and could be signs of another infection.

LYMPHOMAS

Cancers of the lymphatic system, which refers to the immune cells of the spleen, bone marrow, thymus, and lymph nodes are called *lymphomas*. The two primary kinds of this type of cancer are Hodgkin's lymphoma and non-

Hodgkin's lymphoma. Both kinds of lymphomas can occur in adults and children. Like many other types of cancer, the prognosis and treatment are different depending on the diagnosis and the stage of the cancer.

Hodgkin's Lymphoma

Like all other cancers of the lymphatic system, Hodgkin's lymphoma (which is named after Dr. Thomas Hodgkin, who identified the disease in 1932) begins in the body's lymphatic tissue, where lymph nodes produce lymphocytes. In 2002, an estimated 7,000 new cases of Hodgkin's disease were diagnosed in the United States, according to statistics from the American Cancer Society. The disease affects more men than women, but can occur in both children and adults. Hodgkin's disease is most often diagnosed in early adulthood (ages 25 to 30) and late adulthood (after age 55).

This type of cancer can begin growing anywhere in the body, because lymphatic tissue is present in so many regions of the body. In most cases, however, Hodgkin's disease starts in the lymph nodes located in the upper part of the body—in the chest, neck, or under the arms. This disease primarily spreads through the lymphatic vessels, which connect the lymph node network, enlarging the lymph nodes as they become infected. As the lymph nodes grow, they can put pressure on important organs in the body.

In order for patients to be diagnosed with Hodgkin's disease, doctors must determine the presence of cancer cells called *Reed-Sternberg cells*. Scientists have found that these cells look different from cancer cells associated with non-Hodgkin's lymphoma and other cancers. Reed-Sternberg cells have been identified by researchers as a type of malignant lymphocyte that contain two or more nuclei. Some scientists have even noted their unique shape, which causes them to look like "owl's eyes" under a microscope.

While researchers have not determined a specific cause of Hodgkin's disease, this type of cancer has been linked to some risk factors, although it is important to note that most patients are diagnosed without any risk factors. Scientists have found that there is a slight increase in the rate of Hodgkin's disease in people who have had infectious mononucleosis (also known as "mono"), which is an infection caused by the Epstein-Barr virus.

Chemotherapy and/or radiation therapy have been found to be the most effective forms of treatment against the disease. Two common drug combinations used in chemotherapy treatment against Hodgkin's disease are MVPP (mechlorethamine, vincristine, procarbazine, and prednisone) and ABVD (Adriamycin, bleomycin, vinblastine, and dacarbazine). According to American Cancer Society statistics, the survival rate after one year of treatment is 91 percent, while the five-year rate is 82 percent and the ten-year survival rate is 73 percent.

In some cases, however, patients are resistant to standard treatments. Two newer treatments are autologous bone marrow transplant and peripheral blood

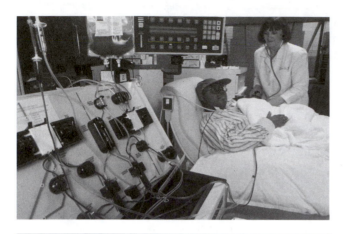

Cancer patient undergoing stem cell collection in preparation for high-dose chemotherapy. © K. Beebe/Custom Medical Stock Photo.

stem cell transplant. In the bone marrow transplant, doctors remove and freeze a portion of the patient's own bone marrow. Then, high doses of chemotherapy (and sometimes radiation) are given to the patient, which not only kills the cancer cells but also destroys the bone marrow. Following these treatments, the marrow is thawed and injected back into the body through a vein. This thawed marrow replaces the marrow that was destroyed throughout the chemotherapy treatments, and begins to produce the blood cells that carry oxygen throughout the body and fight infection. The second type of treatment, peripheral blood stem cell transplant, is similar to a bone marrow transplant, except that stem cells are used instead of bone marrow. The transplant begins with a procedure called *leukapheresis*, which involves the removal of the patient's blood in small amounts so only the stem cells are extracted. The blood is then frozen, the patient undergoes chemotherapy courses, and then the frozen stem cells are thawed and returned to the body (see "Stem Cells").

Non-Hodgkin's Lymphoma

Lymphomas that do not exhibit the presence of Reed-Sternberg cells are known as non-Hodgkin's lymphomas. Statistics show that approximately 53,900 Americans (28,200 men and 25,700 women) were diagnosed with non-Hodgkin's lymphoma in 2002. This number includes both adults and children, however over 95 percent of cases occur in adults.

There are currently no early detection tests for this type of cancer, but researchers have noted certain signs and symptoms that should be examined by a doctor in case treatment is needed. Many patients, or even family members of patients, will notice palpated lymph nodes, which means lymph nodes that are swollen to the extent that they are apparent through casual observation. These lymph nodes are close to the body's surface, and are often located on the sides of the neck, the groin or underarm areas, or above the collar bone. When the lymph tissue or lymphoid tissue around the abdomen is affected, the area can be swollen so that the stomach resembles that of a pregnant woman, due to a fluid-retaining tumor. The lining of the

Stem Cells

Two factors distinguish stem cells from other types of cells in the human body. First, they are unspecialized, which means they do not yet have a specific function in the body. Secondly, these cells can become specialized or be induced to perform special functions under certain experimental conditions. For example, stem cells can be manipulated to become beating cells in the heart muscle, or insulin-producing cells in the pancreas.

Scientists work with embryonic and adult stem cells from humans and from animals. In the early 1980s, researchers discovered ways to derive stem cells from mouse embryos. Then in 1998, researchers discovered how to derive stem cells from human embryos. These embryos were then used to produce cells in a laboratory setting, and were labeled as human embryonic stem cells.

Stem cells are vital to all living organisms. When the embryo is between three to five days old it is called a **blastocyst**. At this stage, the inner mass of a blastocyst contains a small group of about thirty cells, which produce hundreds of specialized cell types that make up an organism. Stem cells in the tissues of a developing fetus produce multiple specialized cell types that make up the heart, lung, skin, and other tissues. Various living tissues in an adult contain significant populations of stem cells in the bone marrow, muscle, and brain that serve as replacements for those cells lost through daily living, injury, or diseases such as cancer.

Many scientists and researchers believe that in the future, stem cells may be part of important treatments for various disorders of the human body. The development of drug therapies to combat various illnesses, disorders (including birth defects), and toxins might also involve stem cells. However, much remains unknown about stem cells. There are two key areas that researchers are focusing on: (1) discovering how stem cells remain unspecialized and how they renew themselves, and (2) determining how stem cells become specialized. These two areas can lead to two treatment strategies currently at the forefront of stem cell research. The first strategy involves growing differentiated cells in a laboratory setting in order to culture them, or nudge them toward a desired cell type before implantation. Another possibility is to implant them directly into the desired area, such as the brain, and rely on signals inside the body to direct their development into the right kind of brain cell. The second research strategy involves finding specific growth factors, hormones, or other signaling molecules that will help cells to live and develop. Hopefully, these factors or molecules would stimulate the stem cells to produce specialized cells to aid in recovery from disease and injury.

abdominal cavity might be damaged by the cancer, causing fluid retention in the stomach lining. Swelling of lymphoid tissue around the intestines can cause feces passage to be blocked, resulting in bowel discomfort or abdominal pain.

In addition to swelling, many non-Hodgkin's lymphoma patients experience generalized symptoms, also known as B symptoms. These include sudden weight loss, fever, excessive sweating, and severe itchiness. The extent of these B symptoms often indicates a great presence of cancer cells in the body.

It's important to note that many disorders of the lymphatic system cause the swelling of the lymph nodes. Therefore, doctors will often monitor the degree of swelling over several weeks and observe how the system responds to antibiotic treatment before conducting a **biopsy**, which is when a small piece of the node, or even the entire node, may be surgically removed for microscope analysis and other extensive testing to determine the presence of cancer cells. Some medical professionals will decide to pursue an immediate biopsy based on the size, texture, or location of the node in addition to the severity of the B symptoms (see "Biopsies").

There are a wide variety of these types of non-Hodgkin's lymphoma, so it helps doctors to classify the lymphatic tumor based on size, shape, and the growth pattern within the lymph node, which indicates the cell type of the lymphoma. This enables the physician to develop a prognosis and treatment regime.

The size of the lymphatic tumors are described as either large or small, while the shape is described as cleaved (exhibiting folds or indentations) or noncleaved (smooth). In terms of growth patterns, the tumor's cell composition can be described as follicular (cells are clustered) or diffuse (cells appear to be scattered). Not every tumor or lymphoma needs to incorporate every term in order to be described. For instance, lymphomas with small and large tumors that appear to have scattered cell distribution would be described as diffuse mixed cell type.

Another way non-Hodgkin's lymphoma is classified is growth rate. There are three categories:

> *Low grade.* This type of lymphoma tends to grow slowly and, in some cases, does not require immediate treatment. Some patients with low-grade lymphoma can live for years without experiencing any of the negative effects of the disease. Follicular small cleaved cell type and follicular mixed cell type lymphomas often indicate low-grade lymphoma.
>
> *Intermediate grade.* Lymphomas at this stage grow rapidly and need immediate treatment. The cell types are generally larger at this grade; follicular large and diffuse large cells are in this category.
>
> *High grade.* Like the intermediate grade, the lymphoma is growing rapidly if it is classified in this category. Therapies such as chemotherapy and radiation are common treatments for non-Hodgkin's lymphomas in the intermediate and high grades.

Treatment

Like other types of cancer, the treatment depends on the stage of the illness, in addition to the specific kind of cancer. In some cases, the area of the body most affected by the cancer—such as the lymphatic or nervous sys-

Biopsies

A biopsy is when a section of the lymph node (or even the entire lymph node) is removed for additional testing to determine whether cancer cells are present. Because a primary indication of all lymphatic disorders is the swelling of lymph nodes, doctors perform biopsies in order to reach an accurate and comprehensive diagnosis. The following are four types of common biopsy procedures.

Fine needle aspiration biopsy (FNA): This procedure uses a thin needle with a syringe to extract a small amount of tissue from the tumor mass. If the tumor is located near the body's surface, then a doctor will often target the needle simply by feeling the enlarged node. However, sometimes the tumor is deep inside the body. In this instance, the needle can be guided with the help of a CT scan. There is a significant advantage to using this technique because it does not require surgery, although sometimes the needle does not remove enough tissue to determine an accurate diagnosis of cancer, such as non-Hodgkin's lymphoma. FNA is particularly effective at diagnosing the presence of cancer after it has spread to nodes from other organs.

Excisional or incisional biopsy: This is when an entire node or a small section of a large tumor is removed. Unlike the FNA, this biopsy is effective at extracting enough tissue to reach a comprehensive lymphoma diagnosis. If the node or tumor is located near the surface of the skin, the surgery can be done using local anesthesia (numbing medication). If the node is deep inside the chest or abdomen, the patient will be put to sleep using general anesthesia for the operation.

Bone marrow aspiration and biopsy: For the aspiration portion of this procedure, a thin needle and syringe are used to remove small amounts of the bone marrow. The biopsy procedure is more extensive. A larger needle is used to remove a cylinder of the bone marrow (approximately one inch long and one-sixteenth of an inch across) located at the back of the hip. This test is not only used for the initial diagnosis, but is also used to determine the stage of the cancer, or how far it has spread throughout the body.

Lumbar puncture: Also referred to as a spinal tap, this procedure uses a thin needle inserted into the spinal cavity located in the lower back. The fluid in this cavity, called the cerebrospinal fluid, is examined for the presence of cancer cells.

Biopsy Tests

After a biopsy is performed, the specimen is analyzed in order to diagnose the patient with cancer. If the patient is diagnosed with the cancer, the following test will also help doctors to classify the specifics of the cancer. A pathologist (a doctor that specializes in the study of diseases) examines the size and shape of the specimen's cells, including how the cells are arranged in the lymph node. Sometimes, however, these tests do not provide a conclusive answer and more testing is needed.

Immunohistochemistry: Cells are treated with special laboratory antibodies at the beginning of this test. As a result, certain types of cells change color. The cells are analyzed under a microscope in order to distinguish the different types of cancer from one another and other diseases.

Flow cyometry: The cells under examination in this test are also treated with special laboratory antibodies, and are then passed in front of a laser beam. Each antibody sticks to a certain kind of cell, and if the sample contains that cell, the laser will enable them to give off a unique color of light that is measured and analyzed by a computer. Similar to the immunohistochemistry procedure, the flow cyometry procedure is important in determining the exact cause of the lymph node swelling, whether it is lymphoma or some other infection.

---

tem, which would include the brain—is the most important factor in determining the treatment option.

Chemotherapy and radiation are two options, but some other treatments, including stem cell transplantation, are also treatment regimes that doctors now consider. Two of these involve interferons and **monoclonal antibodies**, which are both considered biological therapies, meaning that they use substances that occur naturally in the immune system that may kill lymphoma cells, slow the growth of these cells, or even boost the immune system's ability to fight the lymphoma.

The first biological therapy, interferons, are proteins that behave like hormones and are produced by white blood cells to aid the immune system against infections. Numerous clinical studies have found that providing the body with additional supplies of interferons causes some non-Hodgkin's lymphoma tumors to shrink. Many doctors use interferons in addition to chemotherapy. However, some side effects of the interferon treatment include severe fatigue, fever, chills, headaches, muscle, and joint aches, in addition to mood changes.

The second treatment involves monoclonal antibodies. Although these substances are produced in a laboratory setting, they behave like antibodies, which are normally produced by the immune system to fight infections. These monoclonal antibodies are formulated to attack lymphoma cells, just as the natural antibodies attack germs in the body. One type of monoclonal antibody has been found to be effective for the treatment of follicular lymphoma because it attaches to a substance called CD20, which is found on the surface of some types of non-Hodgkin's lymphoma cells, and subsequently causes the lymphoma cells to die. This treatment shares some of the same side effects as interferons, however, including chills, fever, nausea, and headaches.

AIDS-Related Lymphoma

As detailed in Chapter 9, the human immunodeficiency virus (HIV) and acquired immunodeficiency syndrome (AIDS) were first described in 1981.

Shortly after AIDS first became public, doctors noted the incidence of non-Hodgkin's lymphoma among some patients infected with the HIV and AIDS. Studies have found that in many cases, the AIDS diagnosis precedes the onset of non-Hodgkin's lymphoma, although for some patients, diagnosis of non-Hodgkin's lymphoma comes at the same time as the AIDS and HIV-positive determination. Scientists have also found that HIV-associated Hodgkin's lymphoma is also very aggressive when compared to cases where the HIV virus is not present.

Doctors and researchers have found that non-Hodgkin's lymphoma in AIDS patients is particularly aggressive. The advanced-stage location of the disease is often **extranodal**. These sites include the bone marrow, liver, meninges, and gastrointestinal tract. Unusual sites can include the anus, heart, bile duct, and muscles. In addition to its aggressive nature, the lymphoma is more extensive and also tends to respond poorly to chemotherapy. Deficiencies in the immune system—the hallmark of AIDS and HIV—only exacerbate the chemotherapy administration. This creates a vicious cycle, with chemotherapy and radiation weakening an already depleted immune system resource, while the weakened immune system makes the body less responsive to chemotherapy.

THYMUS CANCER

As stated in Chapter 2, the thymus is the element of the lymphatic system responsible for producing T lymphoctyes, which are an important type of white blood cell that is vital for the immune system to function. After the T lymphocytes are produced in the thymus, they travel through the body's extensive network of lymph nodes, where they assist the immune system in protecting the body from various infectious agents, including germs and bacteria.

In addition to the lymphocytes, the thymus also contains thymic epithelial cells, which can become abnormal cancer cells called *thymomas* and *thymic carcinomas*. As stated above, lymphocytes can also develop into cancer cells, which cause Hodgkin's disease and non-Hodgkin's lymphoma. The thymus also contains a third and much less common type of cell called the *Kulchitsky cell*, which is also known as the neuroendocrine cell. These cells release certain hormones, but they can also give rise to cancer cells called carcinoids or **carcinoid tumors**, which are also hormone producing. Of these three kinds of thymic cancer cells—thymomas, carcinomas, and carcinoid tumors—over 90 percent of the tumors they form are thymomas. Scientists look at these differences and evaluate to what extent these cancer cells are invasive, meaning if they have spread beyond the thymus into other organs, and if they have, how far they have spread.

Symptoms

Some thymic cancer patients experience symptoms for at least six months before seeking medical attention, but many do not experience any symptoms, and are only diagnosed when a chest x-ray or other screening shows evidence of a tumor in or near the thymus. In many cases, thymic tumors usually become apparent when they become large enough to compress air passages or blood vessels. For example, if the thymic tumor has grown so that it is putting pressure on the trachea or wind pipe, the patient will experience shortness of breath. Facial swelling could result from the compression of vessels that return blood from the head and neck to the heart.

There are three tumor-related conditions, called *paraneoplastic syndromes*, that are associated with thymomas, although they are not directly related to the pressure from the tumor itself. These conditions are myasthenia gravis, red cell aplasia, and hypogammaglobulinemia. Myasthenia gravis is a type of **autoimmune disease** which occurs when the immune system produces antibodies that attack normal cells and interfere with the functioning of these cells. When patients have myasthenia gravis, their body produces antibodies that interfere with receptors located on the surface of the muscle fibers. These receptors act to stimulate muscle movement, and when their function is attacked, the muscle weakens. This muscular debilitation is experienced throughout the body, and muscles in the eyes, neck, and chest are especially affected, resulting in double or blurred vision and problems with swallowing.

Red cell aplasia refers to a lack of red cell formation, which leads to anemia and low red blood cell counts. This condition typically occurs in adults older than 40, and treatment often requires removal of the thymus gland. Symptoms associated with red cell aplasia (and anemia), include weakness, dizziness, and shortness of breath. The third disorder, hypogammaglobulinemia, occurs when the body produces a diminished supply of antibodies. Because of the abnormal supply, the patient is left vulnerable to infections.

Other symptoms that can occur in thymomatic patients may indicate that the patient's blood vessels or airways in the chest are compressed. In addition, thymic carcinoids can cause carcinoid syndrome, which is linked to the production of certain hormones released by carcinoid tumors. Symptoms of carcinoid syndrome include flushing (increased redness and warmth of the skin due to increased blood flow), in addition to diarrhea and asthma.

9

Diseases of the Lymphatic System: Primary and Secondary Lymphatic Organs

The body's lymphatic system works as both a support and defense mechanism. The lymphs, in addition to the node and vessel components, support the circulatory system by draining excessive proteins and fluids back into the bloodstream, which prevents tissues from swelling. As a defense mechanism, the lymphatic system produces white blood cells and antibodies, in addition to filtering out other microbes and organisms that could cause disease. Any damage to the primary and secondary organs of the lymphatic systems makes the body more vulnerable to infection, which can vary in severity from tonsillitis to cancer. This chapter will focus on some of the diseases and disorders of the primary and secondary lymphatic organs, with the exception of cancer (see Chapter 8).

ACQUIRED IMMUNODEFICIENCY SYNDROME (AIDS)

Since AIDS, or acquired immunodeficiency syndrome, was initially reported in the United States in 1981, the disease has become a worldwide epidemic. AIDS is caused by the human immunodeficiency virus (HIV), which was first identified in 1982, although scientists believe that the virus entered the American population some time in the late 1970s. Since AIDS was discovered, the U.S. Centers for Disease Control and Prevention (CDC) estimates that over 700,000 cases and over 400,000 deaths have been reported among AIDS patients. The CDC estimates that approximately 40,000 new infections of HIV occur annually in the United States, 70 percent of them among men and 30 percent among women. Minority populations appear to be more at risk for infection; AIDS affects African Americans seven times more and Hispanics three times more than Caucasians.

While AIDS and HIV has been studied for over twenty years, there is still a significant amount that is unknown about this disease. In particular, scientists have yet to find a way to stop HIV from causing AIDS, in addition to being unable to find a cure for AIDS. When someone is infected with HIV, it causes the immune system to gradually deteriorate and eventually shut down. This collapse of immune function leaves the body vulnerable to diseases, such as pneumonia and other **opportunistic infections**, including various forms of cancer, that develop following the impairment of the immune system. Specifically, HIV disables and then kills CD4+ T cells (or T helper cells) that play a vital role in producing an immune response and signaling other immune system cells to protect the body from harmful substances and infections. In a healthy person, the T cell count is usually 800 to 1,200 for every cubic milliliter of blood. When one is infected with HIV, this T cell count gradually declines. Once the T cell count falls below 200 per cubic milliliter, the patient becomes vulnerable to opportunistic infections that are characteristic of AIDS. Infections of the eyes, lungs, brain, and intestines, in addition to severe weight loss and diarrhea, are typical upon the onset of AIDS.

Transmission

The most common way that HIV is transmitted is through body fluids, particularly blood and semen. Having unprotected sex (meaning without a condom) is the primary way HIV is spread. The virus can leave the body through the semen, and enter the partner's body through the mucus linings of the vagina, vulva, penis, rectum, or mouth. Before the transmission of HIV was completely understood, the virus could be spread through transfusions using infected blood. However, all donated blood is now screened for HIV. Needle sharing among drug users who inject drugs such as heroin puts the participants at risk for HIV infection. Once someone infected with HIV uses the needle, his contaminated blood can then be transmitted from the needle into someone else's bloodstream.

Pregnant women who have HIV can transmit the infection to their children before or even during birth. Approximately 25 percent to 30 percent of these infected mothers will pass the disease on to their children (HIV can also be spread through breast milk). However, the risk of transmission is greatly reduced if the infected mother takes a drug called AZT (3'-azido-3'-deoxythymidine) during pregnancy. Studies have shown that if AZT is taken during pregnancy and the baby is delivered through a cesarean section, the baby has a less than 1 percent chance of becoming infected. In addition, the mother's placenta acts as a barrier, which prevents the fetal transmission rate from being 100 percent.

In addition to sharing drug needles and having unprotected sex, research has shown that having a sexually transmitted disease such as syphilis, genital herpes, chlamydia, or gonorrhea increases one's susceptibility to HIV while having unprotected sex with an infected partner. However, research has found no evidence of HIV transmission through kissing, sweat, tears, urine, or feces, although saliva has been found to contain HIV in infected patients. Based on studies of families with an infected member, casual contact such as the sharing of food utensils, towels, bedding, swimming pools, telephones, or toilet seats has been shown not to spread HIV. In addition, mosquitoes or other biting insects, such as bed bugs, have also been shown not to transmit the virus.

Symptoms of HIV and AIDS

Many people do not show symptoms of HIV upon the initial infection, meaning they are **asymptomatic**. For others, symptoms such as a fever, headache, exhaustion, or enlarged lymph nodes often appear within a month or two after infection. While these symptoms usually disappear after two weeks or month, the person is very infectious and a significant amount of HIV is present in the body's genital fluids. The more severe symptoms that characterize the onset of AIDS could lay dormant for ten or more years following the initial infection, although the onset is usually two years in children born HIV-infected. This is what makes this disorder so individualized; the type and timeline of symptom onset is unique in each person. Some may feel sick soon after they are infected, while others do not feel any symptoms of illness for ten or more years.

But even if a patient is asymptomatic, the virus is working to multiply itself, in addition to infecting and killing the cells, especially the T cells, which are one of the immune system's primary fighters against infection. As the virus works to debilitate the immune system, infections begin to take over the body. Often the first sign of infection is the persistent swelling of the lymph nodes. Other infections can include exhaustion, sudden weight loss, fevers, sweats, oral or vaginal yeast infections, skin rashes, and short-term memory loss. Some HIV-infected patients also suffer from **herpes** infections that cause sores in the mouth and genital areas. During this period, the normal growing patterns of an HIV-infected child might be slowed or stunted.

The lymph nodes are an active site for this virus. Research has shown that in the early stages of HIV infection, the individual cells of the virus are busy replicating within the lymph nodes and related organs. A significant portion of the virus becomes trapped in the nodes' follicular dendritic cells (FDCs), which are long, tentacle-like extensions made of specialized cells. These FDCs are located in the lymphatic system's "hot spot," also called the body's germinal centers, which are found in lymphatic organs such as the lymph nodes, tonsils, and spleen. The tissues of the FDCs work to trap invading pathogens, such as HIV, and hold them hostage until the B cells arrive and initiate an immune response.

Once the B cells arrive, the CD4+ T cells follow, rushing into these germinal centers to aid the B cells in their fight against the foreign pathogens. These T cells are HIV's primary target, and they become infected once they are in contact with the HIV that is trapped on the FDCs. Scientists believe that this cellular behavior enables an FDC to be an important HIV reservoir, and may explain the development of HIV and AIDS.

The onset of AIDS signals the most advanced stages of HIV infection. The CDC has developed the definition for AIDS, and consequently is responsible for tracking the spread and scope of the disease in the United States. While healthy adults usually have a T cell count of 1,000, HIV-infected patients are diagnosed with AIDS when their T cell count falls below 200. The definition of AIDS also encompasses twenty-six conditions that have been found to affect those with HIV, namely opportunistic infections such as pneumonia and influenza. People with healthy immune systems can usually recover from these infections. But an AIDS patient with a compromised immune system can be powerless and unable to fend off certain microbes, bacteria, fungi, and other infections that lead to disease.

Children with AIDS are vulnerable to bacterial infections such as conjunctivitis (eye infection), ear infections, and tonsillitis. Adults with AIDS are vulnerable to certain types of cancers, especially those caused by viruses, including Kaposi's sarcoma, cervical cancer, or cancers of the immune system also known as lymphomas (see Chapter 8). Red, brown, or purple skin spots are signs of Kaposi's sarcoma in AIDS patients. These cancers are especially aggressive (and very difficult to treat) in people with AIDS.

Treatment

When a patient is diagnosed today with an HIV infection, there are many more medications available to fight both the infections and related illnesses than there were a decade ago (see "HIV Testing"). When AIDS patients started appearing at doctors' offices in the early 1980s, there was nothing available to treat the foundation of the disease—the deterioration of immune function—and there were only a few medications to treat the opportunistic infections.

The initial group of drugs that the FDA approved to combat HIV infection were called *nucleoside reverse transcriptase (RT) inhibitors*, which stopped the initial form of the virus from copying itself and reproducing. Other drugs in the class include AZT, ddC (zalcitabine), ddI (dideoxyinosine), d4T (stavudine), 3TC (lamivudine), and tenofovir (viread). These RT inhibitors were found to be successful in slowing down the rate that the

HIV Testing

Because the early stages of HIV infection often display no symptoms, the presence of the virus is determined by a blood test. The doctor or healthcare provider will analyze the blood for the presence of HIV antibodies. The presence of HIV antibodies in the bloodstream indicates that the virus has infected the body, thus prompting the immune system to react. While researchers have found that HIV antibody levels are not detectable for one to three months following infection, public health officials recommend that people be tested within six weeks to twelve months after being exposed to the virus.

Early detection serves as an alert to those infected with HIV to avoid high-risk behaviors, such as unprotected sex and needle sharing. In addition, healthcare providers can begin the patient on treatment regimens to boost the immune system in order to avoid opportunistic infections.

virus spread throughout the body, in addition to delaying the onset of opportunistic infections. In recent years, the FDA approved a second group of drugs called *protease inhibitors*, which interrupt the virus replication activity later in its lifecycle. Some of these drugs include Ritonavir (Norvir), Saquinivir (Invirase), Indinavir (Crixivan), and Nelfinavir (Viracept).

Doctors and healthcare providers rely on a combination of therapies to treat HIV, because the virus becomes resistant to the medication. The drug regimen called *highly active anti-retroviral therapy* (HAART) involves RT inhibitors and protease inhibitors being used together. Research has found this therapy to be especially effective in reducing the number of deaths from AIDS in the United States, although it is by no means a cure. HAART has been credited with improving the health of AIDS patients by warding off opportunistic infections because it reduces the amount of virus in the bloodstream.

Although HAART is effective at improving life expectancy, it has some harsh side effects. RT inhibitors can cause a decrease

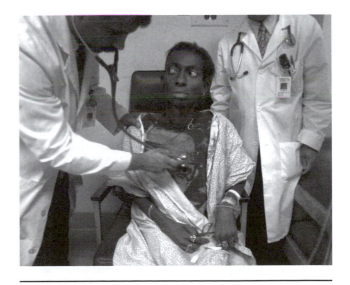

A patient with the AIDS virus is examined by doctors at Brooklyn's Interfaith Medical Center, 1997. © AP/Wide World Photos.

in red or white blood cells in some patients. Some medications have also been found to cause inflammation of the pancreas, in addition to painful nerve damage. Protease inhibitors also have unpleasant to harsh side effects, including nausea and diarrhea, in addition to other gastrointestinal symptoms.

Prevention

Because there still is no cure for HIV and AIDS, health officials continue to focus their efforts on education and related measures to prevent infection. The primary aim of education outreach is informing the public about ways to prevent infection by avoiding risky behaviors, such as sharing needles and having unprotected sex. The only 100 percent sure way to avoid HIV infection through sexual activity is to abstain from sexual intercourse. However, male latex condoms and female polyurethane condoms do offer partial protection during oral, anal, and vaginal intercourse. Latex condoms often feature **spermicide**, which laboratory tests have found can kill HIV. However, there is no evidence to suggest that spermicide can actually prevent HIV infection.

The Quest for an AIDS Vaccine

An estimated 30 million people worldwide are infected with HIV, a majority of whom live in developing countries where preventative measures, such as condoms, in addition to medication for HIV-related opportunistic infections, are not available. The CDC and the NIH, in addition to numerous other American and international health and medical organizations, are working towards developing a vaccine against HIV. This preventative vaccine would instruct the body's immune system to recognize and protect itself against the virus. Scientists hope that development of this vaccine would have a number of possible outcomes, including preventing infections in some or most people exposed to HIV, which would involve preparing one's immune system to block and eliminate the virus. Even if the vaccine is only effective in some populations, it still would have a major impact on the AIDS epidemic.

One country particularly hard hit by the AIDS epidemic is Thailand. In fact, out of an estimated population of 60 million, approximately 800,000 people were believed to be infected with HIV in 2003. Infection rates continue to rise among drug users in the population that share needles. Education, counseling, and access to sterile needles have helped to slow the epidemic, yet thousands continue to be infected with HIV every year through contaminated needles.

To combat this climbing infection rate, the Thailand government has participated in a variety of clinical trials to test HIV vaccines. All vaccine development takes several years of laboratory research and testing in animals before human testing, also known as **clinical trails**, can begin. The clinical testing process involves three phases before the FDA considers approving it for use by the general population:

Phase I: This initial phase involves a small number of healthy volunteers to test the safety and various doses of the vaccine for twelve to eighteen months. For the HIV vaccine, the testing group would be composed of HIV-negative people who are at low risk for infection.

Phase II: This secondary phase tests hundreds of volunteers who are HIV-negative from both high and low risk populations. Testing can last up to two years, and focuses on the safety and immune responses in the body following the administration of the vaccine.

Phase III: This last phase involves thousands of high and low risk volunteers who are HIV-negative. Testing lasts between three and four years and includes the actual effectiveness of the vaccine, in addition to safety.

BLOOD-RELATED DISORDERS

Some diseases related to the lymphatic system are also considered diseases of the circulatory system because of the immunity impairment that can result from blood-related disorders (the majority of these disorders are covered in the Circulatory System volume of this series). Without the body's ability to produce a healthy supply of blood, including normal red blood cells, the immune system cannot function properly. The bone marrow, which produces red blood cells, is important to the lymphatic system because it is one of the sites where one type of lymphocyte—B cells—are produced. Lymphocytes are a vital part of the lymphatic tissue that allows the body to respond to antigens.

The blood-related diseases described in the following sections—anemia, bleeding disorders, hemophilia, and bone marrow disorders—often result from certain disorders of the lymphatic system, such as cancer and AIDS. Red blood cells carry oxygen throughout the body, and the immune system is impacted and weakened if this process is disrupted by a disease.

Anemia

The most common blood disorder, anemia affects more than 3.4 million people in the United States. Women and people with chronic illnesses are especially vulnerable to developing anemia. There are several different types of this blood disorder and each have their own cause, including deficient supplies of iron or vitamins, blood loss, or even a genetic cause. People with anemia often complain of excessive fatigue, but this disorder can also be a sign of a more serious condition, such as cancer or kidney disease.

Anemia is actually not a disease, but is defined as a condition in which the body is not producing enough red blood cells to transport oxygen to the tissues. This oxygen deficiency is caused by a low amount of hemoglobin, the protein in the blood that carries oxygen, causing a decrease in the size and number of red blood cells. The blood of an anemic patient cannot carry an adequate amount of oxygen to the body's tissues and organs, which is necessary for energy production. The body's organs and tissues must have a certain amount of oxygen in order to function properly.

Iron is vital to the body because it enables the red blood cells (RBCs) to carry oxygen. Each RBC contains hemoglobin proteins, which in turn consist of iron particles. These hemoglobin particles accumulate in the body's bone marrow, which is where the RBCs form. The heart will pump blood that is deprived of oxygen to the lungs, where the RBCs' hemoglobin particles bind to the oxygen, and then circulate the oxygen-rich blood throughout the body, delivering it to tissues and organs. In addition, the blood also returns carbon dioxide, which is left over from the energy production process, back to the lungs, where it is exhaled.

The most common and severe type of this blood condition is called iron-deficiency anemia, which is caused by an insufficient amount of iron in the blood stream. If the body does not have enough iron, the bone marrow has difficulty producing the hemoglobin for the RBCs. The body can recycle iron when blood cells die and use it to produce new blood cells. However, if the body loses that blood, then the iron is also lost. Iron-deficiency anemia is common in women of child-bearing age because of the significant blood loss that accompanies menstruation. In fact, anemia affects about 20 percent of women, who can lose 20–40 milligrams of iron every month while menstruating during their reproductive years. Because of the significant loss of iron during menstruation, medical experts recommend that women at risk for developing iron-deficiency anemia eat a well-balanced diet with a steady supply of iron-rich foods, such as spinach, soybeans, and beef. In addition to menstruation, significant blood loss from an ulcer or other medical condition can also lead to iron-deficiency and increase the risk of developing anemia.

While iron is a key component in the production of RBCs, nutrients such as folate and vitamin B_{12} are also necessary (see Chapter 11). A diminished supply of healthy RBCs can be the result of a diet that does not include enough of these nutrients, especially folate, which can put one at risk for developing vitamin-deficiency anemia. Patients with this type of anemia have bone marrow that produces oversized RBCs called **megaloblasts**. People with intestinal disorders and those who are unable to absorb vitamin B_{12} are at risk of developing this type of anemia. In addition, vitamin-deficiency anemia has sometimes been found to occur as a side effect of certain medications, such as oral contraceptives (birth control pills), anti-seizure medications, and some cancer drugs.

Anemia can also be associated with chronic diseases, such as AIDS, cancer (see Chapter 8), and certain liver diseases. These conditions can all interfere with the body's mechanisms for producing red blood cells. Kidney failure also can cause chronic anemia. One of the kidneys' functions is to produce erythropoietin, a hormone that stimulates the bone marrow in red blood cell production. When the kidneys fail, the body has a short supply of erythropoietin, thus leading to a shortage of healthy red blood cells.

Certain cancers and disorders, such as leukemia and **myelodysplasia**, can also cause anemia because of the illness' affect on the body's blood production processes in the bone marrow. In some cases, these processes are mildly affected, while in other cases, the body shuts down to the point where it can no longer produce blood. Another type of anemia, called **sickle cell anemia**, primarily affects people of African or Arabic descent. It's linked to a defective hemoglobin supply and causes the red blood cells to have a sickle, or crescent-like shape. A sickle cell anemia patient has a constantly low supply of red blood cells, because the cells die prematurely. Because of their abnormal shape, these cells can also interfere with circulation by blocking small blood vessels.

The primary symptom of anemia is fatigue, but other signs of the disorder include weakness, pale skin, and a numbness or coldness in the hands or feet. Routine blood exams done during annual checkups usually alert physicians to blood conditions like anemia. If left untreated, anemic patients might experience arrhythmia—an irregular or rapid heartbeat—because the heart must work harder to pump an increased amount of blood due to the lack of oxygen in the red blood cells. Arrhythmia can lead to a heart attack or stroke.

Treatment for anemia is mostly centered on boosting the body's insufficient supplies of iron and other vitamins. For iron deficiency anemia, this means a daily intake of iron supplements, while vitamin deficient anemia patients need frequent injections and supplements of folate (folic acid) and vitamin B_{12}. When a patient with a chronic illness develops anemia, supplements are often ineffective given the disease's effects on the body. In the case of kidney failure, injections of synthetic erythropoietin or blood transfusions (injecting a blood supply filled with healthy red blood cells) may alleviate the fatigue by stimulating red blood cell production. While sickle cell anemia is considered incurable, treatments include pain-relieving drugs that address complications from circulation blockage, and the administration of oxygen, in addition to folic acid and vitamin supplements, which might relieve intense fatigue. These treatments focus on making the patient more comfortable.

Bleeding Disorders

Doctors and health professionals use the term "bleeding disorders" to define a group of medical impairments that cause poor blood clotting and continuous bleeding. Blood clotting, also known as **coagulation**, is the process that controls bleeding by transforming blood from a liquid form into a solid form.

When a wound-causing injury occurs, small cells in the blood called *platelets* accumulate around the wound. In addition to platelets, the blood contains proteins, calcium, and other substances that react to form a clot, which forms a kind of net over the wound. Over the next several days, the clot strengthens and eventually dissolves when the wound is completely healed. When a patient suffers from a bleeding disorder, his blood does not clot properly—or at all—causing continuous bleeding. Symptoms of these disorders not only include excessive bleeding following an injury, but can also include bruising, frequent nose bleeds, and abnormal menstrual bleeding.

Risks associated with these disorders include scarring of the joints, vision loss due to bleeding in the eye, chronic anemia, and even death, which can result from a large amount of blood loss or bleeding into critical areas such as the brain. There are various causes associated with bleeding disorders:

- genetics, such as in the case of von Willebrand's disease
- immune-system related diseases, which can include allergic reactions to medications or reactions to an infection
- cancers, such as leukemia
- bone marrow problems
- disseminated intravascular coagulation, a condition where the body's clotting functions abnormally; this disorder is associated with child bearing and cancer

Hemophilia

There are various factors in the blood that enable it to clot. Hemophilia is a type of bleeding disorder caused by a deficiency in one of the blood clotting factors. There are two types of hemophilia, type A and type B, that are distinguished by the deficient factor. Hemophilia A is known as "classic hemophilia" and accounts for 80 percent of all cases. The main difference between the two types is the clotting factors. For hemophilia A, this clotting substance is known as Factor VIII, and the substance for hemophilia B is known as clotting Factor IX.

Hemophilia is inherited and is passed on through a defective gene on the X chromosome, meaning females are carriers of this trait. Fifty percent of the male offspring of female carriers have the disease and fifty percent of the female offspring are carriers. All female offspring from males with hemophilia are carriers. Because of these inheritance statistics, hemophilia is very rare in women, although they are carriers. About one in 10,000 males are born hemophiliacs (see "Hemophilia: The Royal Disease").

Hemophilia: The Royal Disease

The link between European royalty and the blood disorder hemophilia was first discovered in the 1800s. Queen Victoria, who ruled England between 1837 and 1901, was a carrier of the defective gene that causes the clotting disorder. She passed the gene on to one of her sons, Leopold, who suffered frequent hemorrhages throughout his life that were reported in the *British Medical Journal* in 1868. Leopold had a daughter, Alice, before he passed away at the age of 31. Alice was a carrier and passed the gene on to her son, Viscount Trematon, who also died young of a brain hemorrhage in 1928.

But the genetic spread of the disorder was not simply limited to England, thanks to the intermarriage of royal families from all over Europe as a means of securing friendly relations and diplomacy. Two of Queen Victoria's daughters were hemophilia carriers, and are believed to have passed the disorder on to Spanish, German, and Russian royal families.

One of Queen Victoria's granddaughters, Alexandria, is probably the most famous carrier. When she married Tsar Nicolas of Russia, in 1900, she became the country's tsarina. Because she was a carrier, it is not surprising that her first son, Tsarevic Alexei, was a hemophiliac. His frequent hemorrhages were of great concern to his parents, because there was little known about the disease and even less known about effective treatments at the time. The tsar and tsarina enlisted the help and advice of the best doctors in Russian and Europe, but nothing could alleviate the profuse bleeding and suffering that Alexei was enduring. The royal couple kept this hidden from the public, even the aristocracy, because this young boy was heir to the throne. While the tsar and tsarina were preoccupied with their son's illness, the political situation in Russia was tumultuous and any hint of weakness surrounding the throne could have lead to panic throughout the country.

Then Rasputin appeared. Dressed in heavy black robes like a holy man, this strange man with hypnotic eyes claimed in be a monk from Siberia. By hypnotizing little Alexei, he seemed to be able to make the bleeding and pain diminish. Relieved that her son's pain was somewhat alleviated, Alexandra became quite attached to Rasputin, pledging undying gratitude for helping little Alexei.

Even though he posed as a holy monk, Rasputin had another, darker side. A great womanizer and drinker, he took advantage of this new influence he had over the Russian leaders. While the tsar and tsarina only saw his healing side, the public soon found out about his true persona; soon the royal couple's reputation was tainted by their reliance on Rasputin. Tensions mounted and the political situation turned more tumultuous as Rasputin gained more influence. This factor contributed to the outbreak of the Russian Revolution in 1917.

Because hemophilia is linked to various gene abnormalities, the severity of the disorder's symptoms is different for every individual, and is determined by what degree the abnormality affects the behavior of the clotting factor. In general, however, symptoms include spontaneous bleeding, bruising, bleeding into joints and associated pain and swelling, hemorrhaging in the gastrointestinal and urinary tracts, and excessive bleeding from cuts, tooth extractions, and surgery. In most cases, hemophilia is diagnosed early in a patient's life, usually before 18 months of age. Diagnosis often follows excessive bleeding following a minor injury, or excessive bruising. But no matter what the patient's age, any injury can lead to a dangerous medical situation if he or she is a hemophiliac.

For example, severe bruising can signify bleeding into the joints and muscles, which can lead to crippling deformities. If the tongue gets cut, hemophilia can cause it to bleed and swell to where it can inhibit breathing if blocking the airway. Even a minor bump on the head can cause significant bleeding into the brain and skull, putting a patient at risk for brain damage or even death.

TREATMENTS

The primary treatment for hemophilia is injecting the patient with the missing clotting factor. The amount of clotting factor introduced into the patient depends upon the severity of the bleeding, where the bleeding occurs, and the patient's size. For hemophilia A, the standard treatment is desmopressin, which is administered as an injection or nasal spray. To avoid a bleeding emergency situation, hemophiliacs and their families are often taught how to self-administer these factors. Doctors and health officials also advise hemophiliacs to avoid situations that might lead to bleeding problems; using aspirin and other analgesics, such as anti-inflammatory drugs, can cause blood to have a thinner consistency, which can cause even greater bruising.

VON WILLEBRAND DISEASE

A second kind of bleeding disorder is called von Willebrand disease, which is a genetic deficiency of the blood's von Willebrand factor, a protein that affects the function of platelets. A Finnish physician named Erik von Willebrand (1870–1949) first reported on the disorder in 1926.

Platelets accumulate at the site of an injury or wound to form a blood clot. The von Willebrand factor forces the platelets to bind to the areas of the blood vessel that are damaged, which is the first step in treating or "plugging" a blood vessel injury. If there is an insufficient amount of the von Willebrand factor, blood platelets cannot stick to the holes in the blood vessel walls and a clot cannot form. Although bleeding eventually stops, it doesn't stop as quickly as it should.

Symptoms of von Willebrand disease are similar to that of hemophilia, including excessive and prolonged bleeding and bruising. Because von Willebrand disease is inherited, doctors and health providers will often determine a patient's genetic background, and perform a variety of tests involving platelet count. In terms of treatment, desmopressin nasal spray is common, as are factor injections, both of which are similar to treatments for hemophiliacs.

Bone Marrow Diseases

Bone marrow disorders are also known as myelodysplastic syndromes (MDS). While these blood and lymphatic-related disorders are rare, they are potentially fatal and occur when the body begins incorrectly producing white and red blood cells in addition to platelets, which results in the formation of abnormal cells.

As stated earlier in this chapter, red blood cells, which are produced in the bone marrow, carry oxygen throughout the bloodstream. White blood cells help the body fight off infections, while platelets help clot the blood. The bone marrow responds to the needs of the body in the production of platelets and red and white blood cells. For instance, if the body is confronting an infection, the bone marrow will increase production of white blood cells. If an injury occurs, the bone marrow will step up production of platelets to assure adequate blood clotting. Within the bone marrow are stem cells, which are in charge of creating these blood cells. Stem cells allow the bone marrow to produce an adequate amount of these cells and platelets because they are able to produce exact replicas of themselves as needed.

However, someone with MDS has a defect in the stem cells located in the bone marrow, which produce abnormal blood cells that dominate and crowd out the healthy blood cells. This results in the production of defective platelets, in addition to red and white blood cells that are unable to function properly. In terms of risk factors, researchers have yet to determine the cause of MDS in many cases, although some patients have been known to develop the disorder following intensive chemotherapy or radiation for certain types of cancers.

MDS often leads to anemia, given the low blood counts. Therefore, symptoms of the disorder include weakness, fatigue, increased bleeding, bruising, and frequent infections. In terms of treatment, antibiotics is an important element in order to control infections. In addition, transfusions of red blood cells and platelets might be necessary. Bone marrow transplantation is another option that in some cases actually cures the disease in younger people. However, MDS often affects older people who are not appropriate candidates for bone marrow transplants (see "Bone Marrow Transplants").

Bone Marrow Transplants

In patients under the age of 55, bone marrow transplants are a type of treatment that can potentially cure MDS. The ideal donors for bone marrow transplantation are an identical twin or a sibling who has a perfect match with the patient. If there isn't a family member who is a match, then a search of bone marrow registries can be attempted to find a potential match from an unrelated donor.

In order to transplant bone marrow, the patient's marrow must be completely destroyed through high doses of chemotherapy and/or radiation, in order to kill cells in the present marrow that might cause MDS to return following the transplant or cause the donor's marrow to be rejected. Older patients might not be able to stand the intense radiation therapy, which is one reason why bone marrow transplants are performed in patients under 55. After the radiation therapy, the donor marrow is given through intravenous transfusion to the patient. For every kilogram of the patient's weight, about one tablespoon of donor marrow is injected. Ideally, the body should then begin producing its own healthy blood cells in approximately two to four weeks. However, this production fails to occur in about 5 to 10 percent of transplant recipients. Consequently, radiation therapy followed by a repeat transplant is necessary.

In some cases, the patient will develop **graft-versus-host disease (GVHD)**, where the new marrow will react in the patient's body. The risk of GVHD is greater among older patients and could be a symptom of a mismatched transplant. The severity of the disease could be mild or life threatening, and can be treated with medication or by the removal of T cells, or T lymphocytes, from the donor marrow. Immune system suppression treatments may also be attempted.

CHRONIC FATIGUE SYNDROME

Chronic fatigue syndrome (CFS) affects approximately 800,000 patients in the United States, and is considered a lymphatic disorder because it is characterized by a dysfunctioning immune system that leads to persistent, unrelenting exhaustion. CFS is defined by the Centers for Disease Control as unexplained, but clinically diagnosed as persistent or relapsing chronic fatigue that is of new or definite onset (meaning that it is not a condition the patient has had since birth), and the disorder afflicts about three times as many women as men. Unfortunately, researchers have failed to pinpoint a cause of CFS, although current research focuses on infectious agents, possible immunologic dysfunctions, and nutritional deficiency. Past research linked CFS to the Epstein-Barr virus, a herpes-like virus that causes infectious mononucleosis. It's now believed that the disorder is not caused exclusively by one infectious agent. Current research continues to investigate possible viral causes. Additionally, cofactors such as genetic predisposition, stress, environment, gender, age, and prior illness appear to play an important role in the development and course of the syndrome.

Many people with CFS complain of flu-like symptoms: sore throat, severe headaches, fever, body aches, and weakness, along with swollen lymph nodes. CFS is not the result of ongoing exertion, nor is it significantly alleviated by rest, and it results in a substantial reduction in previous levels of occupational, educational, social, or personal activities. To be diagnosed with CFS, the patient must have four of the following symptoms: a substantial degree of diminished short-term memory or concentration, sore throat, tender lymph nodes, muscle pain, multi-joint pain without swelling or redness, severe headaches, and an inability to sleep to the point of refreshment.

Many medical observers have noted that CFS often seems to be "triggered" by a stressful event, but in all likelihood the condition was previously dormant. Some people appear to get CFS following a viral infection, a head injury, surgery, excessive use of antibiotics, or a traumatic event, yet it's unlikely that these events could be a primary cause on their own. Despite ongoing clinical research, there is currently no cure or specific treatment for CFS. In addition to the numbing fatigue (for which there is no medication), primary symptoms are often pain, especially in various joints and muscle groups in the back. Medical treatment is limited to alleviating the symptoms, usually with nonsteroidal anti-inflammatory drugs such as ibuprofen. Nonsedating antihistamines also may help to relieve any prominent allergic symptoms, such as a runny nose.

Other common treatments for CFS include low-dose tricyclic agents to improve sleep and relieve mild pain, and antidepressants have been used to treat the inevitable depression of CFS sufferers. Treatments used to ward off symptoms have included experimental drugs such as gamma globulin, as well as dietary supplements and herbal preparations like vitamins B_{12}, C, and A; borage seed oil; and bromelain. Because of a lack of proven effective treatment for CFS, some health professionals advise maintaining good health by eating a balanced diet, getting adequate rest, and exercising regularly.

ELEPHANTITIS

Extreme swelling in the arms and legs are the most visible effects of a lymphatic filariasis, or **elephantitis**. There are more than 120 million people affected by this disease worldwide, and over 40 million are disfigured and incapacitated, according to the World Health Organization (WHO). One-third of those infected live in India, one-third in Africa, and most of the remaining patients live in South Asia, the Western Pacific, and parts of the Americas. From a global standpoint, the WHO claims that elephantitis is the second leading cause of permanent and long-term disability, because the deformity resulting from the mutilation of the limbs and even the genitals causes not only physical crippling, but also serious psycho-social damage.

The disease is caused by *Wuchereria bancrofti* and *Brugia malayi*, parasitic filarial worms that live almost exclusively in humans. These worms settle in the body's lymphatic system, infecting the immune system by disrupting the network of lymph nodes and vessels that maintain the fluid balance between the body's blood and tissues. They live in the body for approximately four to six years, laying millions of minute **larvae** that circulate in the blood.

Transmission occurs through mosquitoes, who bite infected individuals and pick up the worm larvae, also called *microfilariae,* that develop into the viral or infective stage within seven to twenty-one days. The microfilariae then make their way to the mosquitoes' mouth and biting parts, where they are injected into the blood of the next person it bites, reproducing and spreading throughout the bloodstream. The swelling occurs as the parasites accumulate in the blood vessels, thus restricting circulation and causing fluid to increase and build up in surrounding tissues. Disease symptoms, however, sometimes do not occur until years after infection. But the worst symptoms generally appear in adults, and more often in men. In communities where elephantitis is prevalent, the WHO estimates that 10 to 50 percent of men suffer from genital damage, including hydrocoele, which is when the sacs around the testes become filled with fluid.

Because elephantitis relies on mosquitoes for transmission, the prevalence of infection is continuing to increase in tropical and subtropical areas where the disease is already well-established. According to the WHO, one of the main causes of this increase is the rapid and unplanned growth of cities, which creates ideal breeding sites for mosquitoes. In areas where elephantitis is endemic or widespread, the rate of chronic and acute symptoms tend to develop at an increased rate among newcomers in the local population, because they haven't built up any form of immunity to the disease.

In terms of treatment, the WHO operates according to two courses of action: controlling the transmission of elephantitis in communities, especially where the disease is endemic, and controlling the effects of the disease by treating the individual. Both treatments have a significant public health component, because damaged tissues are susceptible to infections. Therefore, simple hygiene measures in addition to antibiotics and antifungal agents are important in preventing and treating infections.

LYMPHEDEMA

This lymphatic system disorder is characterized by an accumulation of lymphatic fluid in the interstitial tissue, which leads to swelling in the arms and legs, in addition to other parts of the body. Primary lymphedema can occur when lymphatic vessels are missing or impaired, while secondary lymphedema occurs when the lymph nodes are removed. In both cases, the impaired function of the lymphatic system causes the lymphatic fluid to increase so that it exceeds the system's transportation capacity, causing an abnormal amount of fluid to collect in the affected area's tissues. This accumulated, protein-rich fluid interferes with wound healing by reducing the amount of oxygen that is transported to the affected area, in addition to providing an environment where bacteria can grow and develop into **lymphangitis**.

Primary lymphedema can develop before birth, or develop at the onset of puberty. For the most part, the causes of primary lymphedema are unknown, although it has been linked to certain **vascular** disorders. Secondary lymphedema, also known as acquired lymphedema, has been shown to develop following surgery, radiation, infection, or trauma. Surgeries related to cancer, particularly those associated with the breast, prostate, testicles, and the bladder, may require the removal of the lymph nodes, putting patients at the risk of developing secondary lymphedema. This disorder can occur immediately following the operation, or even years later. Chemotherapy and other radiation treatments used to treat cancers and AIDS-related illnesses can damage healthy lymph nodes and vessels, thus increasing the risk for this lymphedema. This damage can cause scar tissue to form on the nodes and vessels, which can interfere with the normal flow of the lymphatic fluid.

Stages of Development

Scientists have identified three stages of development for lymphedema based on an affected area:

Stage 1: Known as the spontaneously reversible stage, the tissue is still at a "pitting" stage; when the area of skin is pressed in by fingertips, it indents and holds the indentation.

Stage 2: During this stage, the skin is spontaneously irreversible, meaning that the skin tissue has a spongy consistency and is "non-pitting"; when pressed by the fingertips, the tissue bounces back without any indentation forming. However, it is during this stage when the affected limbs begin to harden and increase in size.

Stage 3: At the onset of this stage, the affected limb is large and the swelling is irreversible. If left untreated, the lymphatic fluid continues to accumulate in the tissue. In addition, this swollen limb becomes the perfect host to bacteria, putting the patient at risk for infections that can lead to a decrease or loss of function of the limb.

Treatment

The treatment regimen for lymphedema depends on the cause and is specific to each individual's symptoms. If the initial symptoms seem to be caused by an infection (swelling, redness, rash, or blistering may indicate an infection), a health professional will often prescribe antibiotics to combat the infection. If an infection does not appear to be part of the symptoms, the treatment might include draining the affected area, bandaging, or use of compression garments.

SPLEEN DISORDERS

As described in Chapter 3, the spleen is a fist-sized, purplish organ in the lymphatic system that produces, monitors, stores, and destroys blood cells. A primary part of the spleen's role in the body is to produce antibodies that protect the body, along with removing harmful bacteria. Therefore, if the spleen has to be surgically removed (known as a splenectomy), the body's ability to fight infection is diminished, although eventually, other organs such as the liver ramp up their infection-fighting power to compensate for loss of the spleen.

One of the main disorders of this lymphatic organ is related to when the spleen enlarges, which is called *splenomegaly*. The spleen's ability to capture and store blood increases, causing a decrease in the number of red and white blood cells, as well as the platelets, that are circulating throughout the body through the bloodstream. When a splenomegaly occurs, the spleen is trapping large numbers of blood cells, causing the spleen to clog and inevitably disrupting its functioning abilities. Thus begins a vicious cycle: the spleen grows because of the increased blood cells, and because it is enlarged, it is able to trap more blood cells. Depriving the circulatory system of an adequate amount of blood cells can lead to anemia, increased frequency of infections because of a diminished supply of white blood cells, in addition to bleeding problems due to a decreased supply of platelets.

Some causes of splenomegaly include infections such as malaria and tuberculosis, anemia (which can both cause and result from an enlarged spleen), and cirrhosis of the liver, which is linked to alcoholism. Cancers, including Hodgkin's disease, leukemia, and lymphoma can also lead to an enlarged spleen.

In terms of treatment, the doctors will initially attempt to treat the underlying disease that is causing the enlargement. On rare occasions, a splenectomy is necessary, although doctors avoid this because it makes the patient susceptible to serious infections.

Another lymphatic condition results if the spleen ruptures, which is the most common serious complication that occurs from abdominal injury as a result of beatings, car accidents, or athletic injuries. The location of the spleen (the upper left region of the abdomen) makes it vulnerable if a patient sustains a severe blow to the stomach. If the spleen is hit, its covering and inside tissue can tear, causing a large volume of blood to flow from the abdomen. If this occurs, immediate surgery is needed to prevent massive blood loss. Death can occur if so much blood is lost that blood pressure falls, inhibiting oxygen from getting to the brain and heart.

STAPH INFECTIONS

These infections are caused by the *Staphylococcus aureus* (referred to as "staph") bacteria, which is found on the skin and in the nose. In most cases, infections related to staph bacteria are minor skin irritations, such as pimples, boils, and **folliculitis**. Other conditions include scalded skin syndrome and **impetigo**.

Folliculitis occurs when hair follicles become infected. Hair follicles are tiny openings under the skin where strands of hair, or the hair shaft, grows. When someone has folliculitis, small, white pimples form over these follicles at the base of the hair shaft. Sometimes the outside of the pimple becomes red and irritated. If folliculitis is allowed to worsen, a boil, also called a *furuncle*, can form as the staph infection seeps deeper into the hair follicle, affecting the skin's sebaceous, or oil-producing, glands or the subcutaneous, or deeper, tissue under the skin. The effected area of skin will begin to itch and become irritated, and will soon become red and begin to swell. Eventually, a whitish "head" may form. If this head pops or breaks, the boil will drain pus and blood. These boils can crop up anywhere on the skin. In children, the underarms and groin are vulnerable to folliculitis.

Impetigo is a staph-related skin infection that affects the areas of the skin around the nose and mouth. Large blisters filled with cloudy fluid appear in the effected areas. If these blisters break, they ooze this fluid and eventually crust over and become itchy. Preschool and school-age children appear to be vulnerable to impetigo, especially during the summer months.

Folliculitis and impetigo can spread among people (mostly children) through hands and fingernails. However, most boils heal on their own and impetigo is often taken care of with a short course of antibiotics. But staph bacteria can also cause more serious, fatal infections, such as those in the bloodstream and following surgery. While staph infections have traditionally been treated with penicillin, more difficult strains of the bacteria have emerged that are resistant to this medication and other antibiotics.

STREP INFECTIONS

A bacterium called *Group A Streptococcus* (GAS) is found in the throat or on the skin and can lead to mild illnesses, including "strep" throat (which is characterized by a sore, irritated throat), the skin disease impetigo, or more serious diseases related to pneumonia. In many cases, this bacterium can live in a person without even causing illness.

The spread of these bacteria occurs primarily when mucus from the nose or throat of an infected person comes into contact with another person's infected cut or wound. While the spread of infected strep occurs mostly through ill patients, contagiousness is greatly diminished after the infected person has been treated with an antibiotic for at least twenty-four hours.

Even though strep can cause relatively minor infections, more serious illnesses can develop if it infects parts of the body where bacteria is not usually found, including the blood, muscles, or the lungs (see "Pneumonia"). These infections, called *invasive GAS diseases*, include necrotizing fasciitis and toxic shock syndrome. Necrotizing fasciitis, often described as "a flesh-eating bacteria," can slowly destroy skin tissue, fat, and the body's muscles.

Pneumonia

While pneumonia is a lung disease, it has an important lymphatic component; it is caused by a variety of viruses, bacteria, and in some cases, fungi, that the body's immune system cannot fight off. The CDC estimated that approximately 90,000 Americans died in 1999 from one of many types of pneumonia.

Caused by a strain of the strep bacteria known as *Streptococcus pneumoniae*, this illness begins by infecting the upper respiratory regions, and eventually can spread to the blood, lungs, and nervous system. It is primarily spread though coughing, sneezing, or close contact with infected persons, although scientists are presently unsure how the bacteria invades the bloodstream and lungs. Symptoms can onset rapidly and include a high fever, cough, shortness of breath, and chest pains. These symptoms can then lead to nausea, vomiting, headache, exhaustion, and muscle aches.

Treatment is focused on antibiotics, including penicillin, although some types of pneumonia are now emerging that are resisting antibiotics. Therefore, prevention is becoming a priority in the medical community. In February 2000, the U.S. Food and Drug Administration approved a pneumonia vaccine for children, and the CDC recommends that the following people also receive the vaccine:

- adults 65 and older

- patients with chronic illness, such as heart disease and diabetes

- patients with impaired immune systems due to HIV infection or AIDS, lymphoma, leukemia (or other cancers), in addition to those undergoing cancer treatment and those who have recently had an organ or bone marrow transplant

- patients with a damaged spleen or no spleen

In patients with streptococcal toxic shock syndrome (STSS), the infection causes the body's blood pressure to drop rapidly, which causes organs to begin failing. (STSS is not the toxic shock syndrome associated with women using tampons during menstruation.)

In 1999, approximately 9,400 cases of invasive GAS disease were reported in the United States, according to the U.S. Centers for Disease Control and Prevention (CDC). Out of this total, 300 were STSS and 600 were necrotizing fasciitis. So while invasive GAS diseases are not common, people with chronic illnesses, such as cancer, diabetes, and kidney infections, are especially vulnerable if strep works its way into the body.

TONSILLITIS

This inflammation of the pharyngeal tonsils is caused by virus, bacteria, or other immune-related factors. Almost all children in the United States become ill with tonsillitis at least once, although the illness rarely occurs in children under 2 years of age. Instances of tonsillitis caused by the *Streptococcus* bacteria (the same bacteria associated with strep infections) most commonly occur in children ages 5 to 15 years. Bacteria cause 15 to 30 percent of tonsillitis cases, and the strep strain known as *Streptococcus pyogenes* is the underlying culprit in most of these cases.

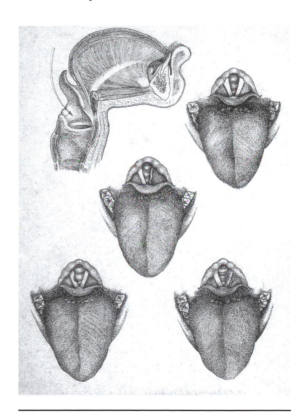

There are four types of tonsillitis: acute, recurrent, chronic, and peritonsillar abscess. Symptoms of acute tonsillitis include a fever, sore throat, dysphagia (difficulty swallowing), odynophagia (painful swallowing), and tender lymph nodes. Because the tonsils are swollen, the patient's airway becomes obstructed, making breathing especially difficult. Symptoms can last from three to four days or up to two weeks. Recurrent tonsillitis is when the patient becomes ill with acute tonsillitis a number of times through the course of a year. Chronic tonsillitis is when patients complain of a constant tenderness of the lymph nodes and a sore throat. Finally, peritonsillar abscess is characterized by se-

Herbert Louis Treush, "Profile of the oral cavity and four vignettes showing the tongue, and normal and inflamed tonsils." © National Library of Medicine.

vere pain in the throat, and can be accompanied by a fever and muffled voice quality.

When tonsillitis is caused by some bacterial factor, antibiotics are usually sufficient for treatment. Because patients might have trouble swallowing, their fluid intake might diminish, putting them at an increased risk of dehydration. In some cases, however, hospitalization is necessary, particularly if the patient's airway is obstructed and normal breathing has become inhibited. In some of these more serious cases, doctors might decide to surgically remove the tonsils.

10

Allergies and
Autoimmune Responses

Allergies and autoimmune reactions are two types of immune-related lymphatic responses. Approximately 50 million people in the United States suffer from some form of allergy-related disease, which can include asthma, food allergies, and hay fever, also known as allergic **rhinitis**. In fact, many doctors and researchers believe that allergic reactions are on the rise among Americans.

Autoimmune-related reactions covered in this chapter are related to the way the body's immune system eliminates bacteria, viruses, and other invading microbes that cause infections (see Chapter 4). If a patient suffers from an autoimmune disease, his immune system functions abnormally so that it attacks the cells, tissues, and organs of the patient's own body.

The first part of this chapter will explore the events behind an allergic reaction, and the different kinds of allergies, and the second part will discuss autoimmune reactions and disorders.

THE ALLERGIC REACTION

As discussed in Chapter 4, the immune system provides a defense against the numerous kinds of bacteria, viruses, and other pathogens that the body encounters simply by breathing, eating, and living. An **allergen** is defined as any substance that triggers an allergic response in the immune system.

An important element of the immune system are antibodies, which float around in the bloodstream and almost all bodily fluids, in an effort to counteract the invading pathogens. One type of antibody is called Immunoglobulin E, or IgE, which reside in mast cells. When an allergen initially enters the body of a allergy-sensitive patient, the immune system perceives the allergen particles as harmful, and then produces an abundant amount of IgE antibodies into the mast cells. Eventually those cells burst and release massive amounts of chemicals and histamines, which set off an avalanche of allergic symptoms. Depending on the patient and type of allergy, these symptoms can affect the skin, causing hives, itching, swelling; the respiratory system, causing swelling of the throat, tightening of the chest, wheezing, shortness of breath; the gastrointestinal system, causing vomiting and diarrhea; and the cardiovascular system, causing a drop in blood pressure, which leads to **anaphylaxis**. It's important to note that each type of IgE antibody has a specific behavior for one particular allergen. For instance, someone who is allergic to cat dander has only IgE antibodies specific to cat dander, while someone allergic to peanuts has IgE antibodies allergic to peanut allergens.

Scientists and researchers are not clear on why some substances cause allergic reactions while others do not. In addition, it is not understood why some people have allergies and others are free to inhale dust, pollen, and cat dander without experiencing any symptoms at all. Research has determined, however, that a family history of allergies is the most important determining factor in developing this disease. Scientists have found that while the average child has a 48 percent chance of developing allergies if one parent has allergies, this risk escalates to 70 percent if both parents have allergies.

TYPES OF ALLERGIC DISEASE

Allergy Rhinitis

Rhinitis, or hay fever, is the most common allergic disease in the United States, and is estimated to affect approximately 40 million people every year. The condition is defined as inflammation of the mucous membranes that line the nose, and its symptoms include sneezing, itching of the nose, watery eyes, in addition to nasal discharge and congestion. While there are some types of rhinitis that are not allergy-related, seasonal and perennial are the two classifications of this disorder that are allergy-related.

Seasonal rhinitis involves the reaction of the body's IgE antibodies located in the nasal mucous membranes to seasonal allergens, which usually come from grass, trees, weed, or fungal (mold) spores. These allergens are more active during the spring and summer when flowers and weeds are in bloom and giving off an allergen known as **pollen**. In patients who have perennial allergic rhinitis, however, symptoms such as nasal blockage and congestion persist through the year. This is because their nasal IgE antibodies are re-

Antihistamines

Histamines are naturally occurring chemicals in the body that are released as a result of an allergic reaction. Antihistamine medications are the primary pharmalogical therapy for many allergic diseases, particularly rhinitis, because they are effective at blocking symptoms associated with the release of histamine, including sneezing and watery eyes, although they have been shown to be generally ineffective against nasal congestion. These drugs also have a rapid onset, which means they quickly begin working to diminish symptoms once they get in the body.

Older antihistamines had a significant sedating effect, which led to fatigue and impaired motor and cognitive abilities. However, nonsedating antihistamines have been available for some time, and do not have these same unsavory side effects.

acting to allergens that do not exhibit any seasonal variation. People with pet allergies have perennial allergic rhinitis, because they have reactions to cat and pet dander. Other common perennial allergens include dust mites or particles, cockroaches, and mold-related substances.

Appropriate treatments for allergic rhinitis include medications such as antihistamines and decongestants to clear out the nasal blockage and congestion. Antihistamines work to block the release of the chemicals that are part of the allergic reaction (see "Antihistamines"). Avoiding factors that trigger an allergic reaction, such as staying away from cats and dogs, is also an important part of the treatment. However, it's practically impossible to escape environmental allergens floating in the air, therefore medication therapies are often necessary.

Asthma

This allergic condition is characterized by respiratory inflammation, which can lead to breathing difficulties if the airway is obstructed. This inflammation is due to an allergic reaction taking place in the airways, which also causes increased bronchial responsiveness and greater sensitivity to allergens. Risk factors for asthma include being sensitive to certain indoor allergens, such as dust mites in the house and animal dander from cats and dogs, in addition to a sensitivity to common outdoor allergens. Researchers have found that exposing children to tobacco smoke at an early age also increases the risk of asthma. Asthma is typically first diagnosed in childhood; 50 to 80 percent of juvenile asthma patients are diagnosed by 5 years of age.

It is common for people who have perennial and seasonal rhinitis to also be asthma sufferers, because both conditions involve a common respiratory mucous membrane. In addition, allergic reactions involving the nasal mucous membranes may cause the lower airways to be more sensitive and responsive to stimuli, a symptom of asthma. Inflammation plays a key role in both asthma and allergic rhinitis. Other symptoms associated with asthma include coughing, wheezing, shortness of breath or rapid breathing due to airway obstruction, and tightness of the chest.

Medical treatments for asthma include use of inhalers, also called beta$_2$-agonists, among other drugs used to treat pulmonary or airway conditions. But an important part of treatment also includes controlling factors that tend to make the asthma more severe, such as avoiding smoky or dusty environments.

Skin Reactions

The broad range of allergic reactions between an external agent and the surface of the skin are known as contact dermatitis. These type of reactions are one of the most common skin diseases in adults, and account for 30 to 40 percent of all workplace-related, or occupational, illnesses. Quite simply, contact dermatitis is a varying degree of inflammation and irritation on the skin that occurs when the immune system perceives a substance (usually a chemical) that comes in to contact with the skin as harmful, thus triggering an allergic reaction (see Table 10.1 for a listing of common skin allergens). While the top of the hands, the eyelids, neck, and genitalia areas tend to be the most reactive to contact dermatitis, the palms of the hands,

TABLE 10.1. Chemicals Associated with Contact Dermatitis

Substance	Found in
Urushiol or Rhus	Poison ivy, poison oak, mango
Saps associated with plants	Philodendron, hydrangea, chrysanthemum, tulip bulbs
Nickel sulfate	Metal alloys, hairpins, earrings, zippers, door handles, hair dyes and bleaches, insecticides
Potassium dichromate	Cement, leather, household cleaners, bleaches
Formaldehyde	Cosmetics, fabrics, cigarettes, newsprint and newspaper, preservatives, cardboard, plywood, rubber cement, shoes
Ethylenediamine	Dyes, fungicides
Mercaptobenzathiazole	Rubber products
Thiuram	Fungicides, insecticides, rubber products
Paraphenylenediamine	Hair dyes, fur dyes, chemicals used in photography

soles of the feet, and the scalp are the most resistant. The reaction can vary from minor redness to large pus-filled vesicles that can become infected.

In most patients, contact dermatitis is localized (confined to where the allergen affects the skin) and usually goes away within three to four weeks. Treatment consists of first minimizing or avoiding contact with the allergen, treating the dermatitis area with a topical corticosteroid medication, along with prescribing an antibiotic if an infection has developed. In addition, patients will be advised to avoid future contact with the allergen.

Latex Reactions

According to the American Academy of Allergy and Immunology (AAAI), an allergy to latex, a substance that is often found in rubber products such as gloves and condoms, is a relatively new, yet increasingly frequent, disorder. While the reasons are unclear, scientists believe the allergy has to do with the increased used of latex products and industry modifications in the processing and manufacturing of these products.

In its initial form, latex appears as a milky fluid produced by rubber trees. Through various different production methods, chemicals are added to latex while it is in a liquid form that increase the speed that the liquid is cured, or vulcanized. This process of vulcanization enables the latex to be impenetrable to oxygen from the air. Products such as rubber bands or car tires that are made either completely from natural rubber latex or from blends of latex and other compounds rarely cause an allergic reaction. However, some people are sensitive to and develop allergic reactions to products that have been dipped in latex, such as balloons, condoms, and latex gloves.

There are four routes of exposure to latex. The skin is exposed to products like gloves; the mucous membranes are exposed through condoms or other medical devices; latex can be inhaled if the person is wearing gloves dusted with powder that contains latex particles; and intravascular exposure can result from medical devices with rubber tips that contain latex. Latex can cause two types of allergic reactions. The first results in contact dermatitis (see previous section), which causes a poison ivy–like rash to appear approximately twelve to thirty-six hours following initial contact. Because this often occurs when people put on latex gloves, the dermatitis appears on the hands, but the rash can spread quickly to other parts of the body. While the contact dermatitis results in an irritating rash, this type of allergic reaction to latex is not life threatening.

The more severe type of reaction to latex is in the form of an immediate, IgE antibody–mediated allergy. This kind of reaction is similar to other severe reactions in that the person does not have a reaction following an initial exposure. But this first exposure causes the person to be sensitized to the latex, which means the reaction will occur upon re-exposure.

The first symptoms of a reaction can include itching, redness, swelling, sneezing, and wheezing in any of the four exposure areas: the skin, mucous membranes, inhalation, and intravascular. In some cases, anaphylaxis and life-threatening symptoms, such as trouble breathing and loss of blood pressure can result. The severity of the latex reaction is dependent on the person's sensitivity to the substance. According to the AAAI, the most severe reactions seem to occur following exposure to moist areas of the body, such as when the lips are exposed to latex. Exposure to these areas of the body allow the allergen to be more rapidly absorbed into the body.

As mentioned earlier, airborne latex particles can also cause severe respiratory problems. In the case of latex gloves, the allergen can attach itself to the cornstarch powder that is often added to the gloves to increase absorption. When the gloves are used, the starch particles, in addition to the allergen, become airborne, and can then be inhaled or come into contact with the nose and eyes. Certain areas of a hospital, such as intensive care units and operating rooms, can have high concentrations of airborne latex, simply because hundreds and hundreds of pairs of these gloves are used everyday.

The AAAI estimates that approximately 1 percent of Americans have a latex allergy, although scientists and doctors have found that certain groups are particularly vulnerable to a reaction, and most of the time this is due to frequent exposure. For instance, patients with spina bifida (a congenital development disorder involving the spinal column) have been found to have a 50 percent risk of developing this allergy. Patients with congenital disorders of the urinary tract have also been found to have approximately a 50 percent risk for developing a latex allergy. In addition, 10 to 17 percent of healthcare workers are believed to have latex allergies, presumably due to frequent exposure.

The AAAI recommends substituting vinyl for latex gloves, although vinyl gloves may not be appropriate in all medical settings and situations. Synthetic latex gloves are also an option, although they are more expensive. However, these synthetic gloves work in nearly every setting where standard latex gloves work, including surgery. For people allergic to latex condoms, those made with natural skin do not contain latex. However, natural skin condoms do not protect against the transmission of HIV, which causes AIDS.

Food Allergies

These types of allergies involve immune-related responses to specific proteins in food. Milk, eggs, peanuts, wheat, soy, and tree nuts (such as cashews, pecans, and walnuts) account for 90 percent of food allergy reactions in children, while peanuts, fish, tree nuts, and shellfish (shrimp, lobster) account for 90 percent of food allergies in adults. Researchers have found that the prevalence of food allergies is greatest during the initial years

TABLE 10.2. Major Allergens in Food

Protein	Food
Caseins, whey	Milk
Ovomucoid	Egg whites
Tropomycin	Shellfish

of a child's life, but this prevalence also declines over the first ten years. Many infants actually outgrow their allergies to eggs, milk, and soy, although food allergies to peanuts, tree nuts, shellfish, and fish are usually not outgrown (see Table 10.2 for some identified proteins that are known to be food allergens).

Food allergies are characterized by severe life-threatening reactions, called *anaphylaxis*. Symptoms of the reactions tend to be particularly rapid when the food is ingested or the allergic patient is exposed to the allergen. A reaction can occur within minutes of exposure and can include some or all of the following symptoms: itching or tingling of the lips, tongue, or throat; swelling of the lips or tongue; feeling of tightness in the throat; nausea or vomiting; abdominal cramps; and diarrhea. An anaphylaxis reaction, or anaphylactic shock, is especially dangerous because it causes a sharp drop in blood pressure. Once the body is in anaphylactic shock, the heart no longer has the power to pump enough blood to the body's major organs, such as the brain, heart, liver, and lungs. Death results when these organs shut down from lack of blood. Severe respiratory reactions can also lead to death, because they cause the throat, tongue, and larynx to swell up, closing the upper airways so the victim chokes to death.

There is no cure for a food allergy, but as noted above, some can even be outgrown. But food allergies force patients to adhere to a diet that totally avoids exposure to the allergen, and that includes becoming educated about the contents of some foods that might unexpectedly contain allergens. For example, food companies have revealed that peanut flour is often used as a thickener in chili, gravy, and even spaghetti sauce. A strict elimination diet, free of the food allergen, is currently the only proven therapy. In case of an accidental exposure to the food allergen, **epinephrine** (adrenaline) should be administered through a pen-like device because it quickly reduces the swelling and prevents the throat from closing shut. In addition, many doctors will recommend using an antihistamine to counteract the swelling. Like food allergies, some insect bites can cause anaphylactic reactions (see "Insect Bites").

Insect Bites

Biting insects, such as honeybees, bumblebees, yellow jackets, wasps, hornets, and fire ants, often cause some degree of swelling and inflammation once they bite or sting, but occasionally they do cause anaphylactic reactions. Other insects, such as mosquitoes, ticks, and some spiders, often cause milder reactions, such as an itchy bump on the skin. Symptoms of a reaction are caused by the insect injecting venom or other allergen substances into the skin. This venom can trigger a reaction. The severity of the reaction depends on a person's sensitivity to the insect venom or other allergen.

Most reactions are mild and cause some irritating itching or stinging that will disappear in a day or two. More serious symptoms might be delayed, such as a fever, pain in the joints, hives, and swollen glands. Symptoms of a severe reaction include facial swelling and difficulty breathing. Epinephrine and antihistamines, in addition to compression and the application of ice, are also used to try to counteract the severe symptoms of insect bite reactions.

AUTOIMMUNE DISEASES AND DISORDERS

While many types of autoimmune disorders are rare, millions of Americans suffer from some type, or from a disease that has an autoimmune damaging component. Women are more likely to suffer from autoimmune disorders, in particular, working women and those in their childbearing years have been found to be more affected by abnormal immune systems. Autoimmune disorders have been found to be more prevalent in certain minority populations. For example, lupus has been found to be more common in African American and Hispanic women than in Caucasian women. Rheumatoid arthritis and scleroderma have been found in greater incidence in certain Native American communities in comparison to the general American population (see Table 10.3 for a listing of some autoimmune diseases).

It's important to note that autoimmune diseases have not been found to be contagious, meaning that they do not spread to others like a cold or infections. In addition, they are not related to AIDS (acquired immune deficiency syndrome), nor are they related to any type of cancer. Research has found a certain genetic component to autoimmune diseases in that the genes inherited might make one susceptible to an autoimmune disease that afflicts one or both his parents. For instance, it is not uncommon for certain diseases, such as psoriasis, to afflict several members of the same family, which indicates that there is some sort of genetic relationship. Certain viral and environmental factors have been known to trigger or worsen an autoimmune disease. In the case of lupus, sunlight can cause the disease to begin and also make it worse. Other influences that affect these types of disease are aging, chronic stress, hormones, and pregnancy.

TABLE 10.3. Autoimmune Diseases

Human Body System and Organs	Disease
Nervous system	Multiple sclerosis, myasthenia gravis, autoimmune neuropathies (such as Guillain-Barré), autoimmune ureitis
Gastrointestinal system	Crohn's disease, ulcerative colitis, primary biliary cirrhosis, autoimmune hepatitis
Blood	Autoimmune hemolytic anemia, pernicious anemia
Blood vessels	Temporal artertis, antiphospholipid syndrome, vasculitides (such as Wegener's granulomatosis), Behçet's disease
Endocrine glands	Type I or immune-mediated diabetes, Grave's disease, autoimmune disease of the adrenal gland
Skin	Psoriasis, dermatitis, pemphigus vulgaris, vitiligo
Musculoskeletal systems	Rheumatoid arthritis, lupus, scleroderma, polymyositis, Sjogren's syndrome

Autoimmunity and the Immune System

One of the basic characteristics of the immune system is that it distinguishes invading antigens as infectious or safe (which are known as self antigens). As explained in Chapters 1 and 4, macrophages and neutrophils are white blood cells that circulate in the body's blood system, looking for harmful antigens. Upon finding these antigens, they engulf and destroy them through the production of toxic molecules known as reactive oxygen intermediate molecules. However, when the immune system is not functioning properly, the production of these molecules can go into overdrive, causing not only the foreign antigens to be destroyed, but also the tissues surrounding the macrophages and neutrophils.

One example is the autoimmune disease called *Wegener's granulomatosis*. When patients have this disease, they have overactive macrophages and neutrophils that invade blood vessels, producing a flood of toxic molecules that cause damage to blood vessels. Another example is rheumatoid arthritis, which also causes macrophages and neutrophils to produce toxic molecules that invade the joints, causing inflammation and possibly permanently damaging these joints.

The B cells are also an important factor in autoimmune disorders. In a normal immune system, these cells come into contact with a specific antigen and become plasma cells, which produce antibodies that are released into the blood's circulation. However, in some autoimmune diseases, the B cells make autoantibodies, which target the body's own tissues rather than foreign antigens. These autoantibodies can either interfere with the normal functions of these tissues or even work to destroy the tissue. For example, when a patient suffers from the skin disease *pemphigus vulgaris*, autoantibodies attack skin cells, causing an accumulation that causes other molecules and cells to break down, eventually causing skin blisters.

When the immune system is especially active, an immune complex can develop. This is when a large amount of antibodies are bound to antigens in the bloodstream, forming a type of connected network throughout the body. This can be harmful when the accumulation leads to inflammation, especially within small blood vessels. All of these immune complexes, cells, and inflammatory molecules can ultimately block up blood vessels, halting blood flow, and eventually destroy organs, such as kidneys.

In order to remove these immune complexes, the body's immunity is equipped with a complement system. This network is made up of various molecules that are found in the bloodstream and on the surface of cells that make the immune complexes more soluble, which reduce their size so they do not accumulate and block blood flow.

DIAGNOSIS

Doctors diagnosis patients with autoimmune disorders based on their symptoms, physical examinations, and laboratory test results. However, these types of diseases can be hard to diagnose because many of the symptoms, such as exhaustion, are symptoms of other disorders. But the sooner doctors pinpoint the exact autoimmune disorder, the better. An early diagnosis allows for aggressive medical therapy. In some cases, patients will respond completely to targeted therapies if the treatment is initiated soon after the onset of symptoms.

One of the hardest aspects of autoimmune diseases is that they tend to follow unpredictable behavior patterns. Doctors cannot determine a disease's course based on initial symptoms, even though the illness is chronic. For this reason, doctors will often monitor patients closely, including analyzing various factors of their patients' life, particularly because environmental factors can trigger or worsen many autoimmune disorders. Patients with autoimmune disorders often have to make frequent visits to their doctors for close scrutinization of medical therapies.

TREATMENT

Because autoimmune diseases are chronic, treatment and monitoring must become part of a patient's life and, in many cases, their daily routine. While many patients can live normal lives under medical supervision and care, few can be cured or completely recover, even through treatment.

In some autoimmune diseases, inflammation is the primary factor that needs to be controlled. For Type 1 diabetes, patients take **insulin** so that blood sugar levels do not elevate and lead to inflammation, which can cause damage to the kidneys, eyes (blindness), blood vessels, and nerves. Much scientific research for this type of diabetes is currently focused on preventing inflammation from destroying the cells in the pancreas that produce insulin, because these cells are needed to control blood sugar levels.

In some types of autoimmune diseases, the abnormal functioning of the immune system slowly erodes the kidneys and joints, as is the case with lupus or rheumatoid arthritis. Medications or therapies that suppress the immune response (immunosuppressive) can help diminish harmful inflammation that causes this erosion. However, some of these medications, such as corticosterioids, azathioprine, and cyclosporin, also have adverse side effects and weaken the immune system against fighting infections. Immunosuppressive medications can lead to remission or disappearance of the disease for a significant amount of time. But even though the disease appears to have left the patient, medication must be continued. Research in the immunosuppressive medication arena is now looking at ways to eliminate various side effects through targeting certain steps of the immune response. Scientists and researchers hope that therapies such as targeting antibodies against specific T cells may produce fewer side effects without sacrificing the effectiveness of current medications.

Specific Autoimmune Disorders

A number of autoimmune diseases are considered rheumatic, meaning that they are characterized by swelling and inflammation. In addition, rheumatic diseases can lead to the loss of function of one or more of the body's connecting or supporting structures, including joints, tendons, ligaments, bones, and muscles. There are more than 100 rheumatic diseases, most of which have common symptoms of swelling, pain, and stiffness.

In rheumatoid arthritis patients, the immune system acts to erode the lining, called the **synovium**, that covers various joints in the body. The disease causes inflammation of the synovium on both sides of the body, in addition to stiffness of the joints. Current therapies include anti-inflammatory drugs to diminish swelling in the joints, in addition to immunosuppressive therapies.

BEHÇET'S DISEASE

First described in 1937 by Istanbul professor of dermatology Dr. Helusi Behçet, this disorder is a chronic skin condition characterized by sores or ulcers in the mouth and in the genital regions. In addition, Behçet's Disease

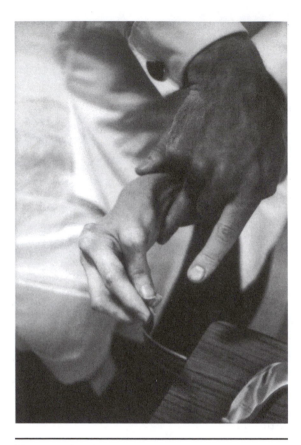

Rheumatoid arthritis: a therapist trying to restore the use of a patient's rheumatic fingers. © National Library of Medicine.

(BD) can also lead to arthritis in the digestive tract, brain, and spinal cord. Although rare in the United States, BD is common in Middle Eastern and Asian countries, where it affects more men than women (the opposite holds true for American patients). BD tends to appear in patients in their late 20s or 30s, although it has been shown to affect all ages.

While the exact cause of BD has not been determined, the primary symptoms are associated with an autoimmune reaction that leads to inflammation of the blood vessels and veins. The disorder is not contagious or spread to other people like an infection, although researchers do believe that there might be a genetic component to BD, although the exact gene or genes have not been identified. Another factor important in the development of this disease appears to be an environmental bacterium or virus that might trigger the disease in those with a susceptibility. Scientists have also discovered that patients who frequently come down with strep throat (see Chapter 9) are more likely to develop this condition.

BD affects each patient differently, and symptoms vary depending on the patient. The four most common symptoms are mouth and genital sores, eye inflammation (in addition to swelling in the visual region), and arthritis. While severe symptoms usually emerge months or years after the initial signs are recognized, some patients only suffer the mild symptoms, including skin sores or ulcers on the mouth or in the genital regions. More severe symptoms can include **meningitis**, which refers to a fatal condition caused by swelling of the membranes that cover and protect the brain and spinal cord. Some symptoms of meningitis include persistent fever and headaches. Symptom outbreaks are known as flares, and can appear and disappear in cycles lasting months or just days. A more serious symptom that affects approximately 10 percent of BD patients is **thrombophlebitis**, which is inflammation of a vein following an obstruction, usually in the leg. This inflammation can lead to more blood clots, and some patients also experience problems with their arteries, such as **aneurysms**. In rare incidences, BD can cause swelling and inflammation in the body's digestive tract, which can cause stomach pain, diarrhea, constipation, and vomiting.

Treatment for BD usually focuses on the symptoms, in addition to corticosteriods and immunosuppressive drugs. Anesthetic topical medications can help alleviate sores and reduce inflammation, but immunosuppressive drugs are also important because they diminish the reactivity of the immune system that is triggering the reaction.

LUPUS

Systematic lupus erythematosus, or lupus, is an autoimmune disease that affects many different parts of the body—from the joints, skin, and kidneys to the heart, lungs, blood vessels, and brain. Common symptoms of the disorder include extreme fatigue or exhaustion, arthritis due to painful and swollen joints, persistent fever, skin rashes, and abnormal kidney functions. While there is currently no cure for lupus, most patients can lead active lives by controlling the severe periods of illness, called *flares*, through medication under a doctor's supervision. Lupus is more common in women, and more likely to affect minority women. In fact, African American women are three times as likely as Caucasian women to develop lupus.

One of the most frustrating aspects of lupus is that researchers have yet to uncover a cause of the disease, although studies indicate that the disease is caused by a combination of genetic, environmental, and hormonal factors. While scientists claim there could be a genetic link, a specific "lupus gene" has yet to be identified. Each patient's experience with lupus is unique; some experience headaches, dizziness, depression, and seizures, while other patients experience hair loss, and anemia. In some patients, only one system of the body is affected, thus impairing the body in different ways (see Table 10.4 for a breakdown of lupus symptoms by body system).

Diagnosing and treating lupus patients is a challenge due to the wide range of symptoms. A team of specialized doctors will usually treat these patients, including immunologists (those specializing in immune system disorders) and rheumatologists (those specializing in rheumatic diseases). Treatment courses primarily include focusing on the patient's symptoms so they can lead an active, productive life.

PSORIASIS

This chronic skin disease has only recently been classified as an immune disorder. Researchers believe that the disorder is associated with T cell activity (or overactivity) in the skin. Psoriasis is characterized by inflamed skin that is scaling—the skin is flaking because cells in the outer layer are reproducing at such a fast rate that they accumulate on the skin. While the disorder mostly affects adults, it occurs about equally in men and women, and approximately between 1 and 2 percent of the population suffers from

TABLE 10.4. Symptoms of Lupus by Body System

System	Effect
Kidneys	Kidneys are important in ridding the body of waste products and toxins. Lupus can cause kidneys to swell and become inflamed, thus causing function to be impaired.
Lungs	Lupus patients can develop pleuritis, a lung disease that causes chest pains and breathing difficulty because of inflammation of the chest cavity.
Central nervous system	The brain and other organs of the nervous system are impaired by lupus, which can lead to headaches, dizziness, memory problems, or even a stroke.
Blood vessels	Lupus can lead to vasculitis, which is inflammation of the blood vessels. This impairs blood circulation throughout the body.
Blood	Anemia and an increased risk of blood clots are associated with lupus.
Heart	Inflammation in the heart or surrounding membranes due to lupus can cause chest pain and other cardiovascular problems.

psoriasis. Psoriasis can occur in thick, red patches on the elbows, knees, scalp, lower back, face, palms, and soles of the feet. Psoriasis can also lead to joint inflammation and arthritis.

Diagnosing psoriasis is difficult because the redness and scaling on the skin can often be mistaken for other skin diseases. However, there are five primary types of psoriasis, the most common kind being plaque psoriasis, which is characterized by a reddened base with scaling. In patients with guttate psoriasis, patients have small sores in the shape of drops on their trunk, limbs, and scalp, while patients with pustular psoriasis have blisters filled with noninfectious pus. Inverse psoriasis is characterized by large and dry red patches located near the genitals, breasts, or in the armpits. Finally, erythrodermic psoriasis features itchy and scaly skin in wide areas, often associated with sunburn or the use of oral steroids.

Treatment is tailored to the type of psoriasis the patient is suffering from, but most courses follow a three-step approach. Initially, topical treatment to relieve the discomfort is applied to the skin. This is then followed by exposure to sunlight, which researchers have found to clear up the irritation in many patients. Finally, immunosuppressive medications and antibiotics are used to supplement treatment.

SCLERODERMA

This autoimmune skin disorder is named from the Greek words "sclerosis," which means hardness, and "derma," which means skin. Therefore, scleroderma literally means "hard skin." There are many different types of scleroderma, but all involve the abnormal growth of the connective tissue that supports the skin and the body's internal organs. Researchers believe that patients with scleroderma have an immune system that produces too much collagen through the stimulation of cells called *fibroblasts*. This collagen then develops into connective tissue that surrounds the cells of the skin and internal organs. In patients with mild scleroderma, the buildup is centered around the skin and blood vessels, while more serious forms of this disorder interfere with functioning of the joints and internal organs.

There are two main types of scleroderma: localized and systemic. While localized diseases affect only certain areas of the body, systemic diseases affect the entire body. Localized scleroderma is limited to the skin, and does not involve any internal organs, such as the kidneys or heart. In many cases, localized scleroderma can heal and even disappear over time, although permanent damage can be left behind. Morphea is one type of localized scleroderma that refers to reddish, thick, oval-shaped patches that most often appear on the chest, stomach, and back, although they can sometimes crop up on the face, arms, and legs. Morphea can sometimes occur as a result of radiation treatment for cancer. Linear scleroderma is characterized by a single band of thickened skin that often appears on the arms or legs.

Systemic scleroderma involves the skin, in addition to the blood vessels and internal organs. This classification is further broken down into limited and diffuse. Limited affects only certain areas of the skin, such as the fingers, hands, face, lower arms, and legs. The condition comes on gradually also, which means it can take a few years before the skin begins to thicken. Diffuse scleroderma has more of a sudden onset, meaning that skin thickening comes on quickly and affects the hands, face, upper arms, upper legs, chest, and stomach. On an internal level, it can damage the heart, lungs, and kidneys. Patients with this type of scleroderma suffer more serious health consequences because of the effect the disease has on these internal organs.

Treatment for scleroderma is similar to psoriasis in that it is typically individualized and based on the patient's symptoms. For instance, a dermatologist might be able to treat the skin irritations, while a nephrologist may be able to treat the kidneys, in case there are internal organ complications. In some cases, scleroderma patients experience thickening of the skin in and around the mouth, which might be able to be treated by a dentist or orthodontist. Currently, however, there is no treatment that cures or even stops the overproduction of collagen, which is the foundation of scleroderma. Therefore, many doctors focus on alleviating symptoms and minimizing damage to internal organs. With all types of autoimmune disorders,

early diagnosis and careful monitoring by a medical professional are para-
mount.

SJOGREN'S SYNDROME

Approximately 1 to 4 million Americans are affected by this autoimmune
disorder, which is characterized by two symptoms: dry eyes and a dry
mouth. Sjogren's syndrome is an autoimmune disease that often accompanies autoimmune conditions such as rheumatoid arthritis, systemic lupus
erythematosus, and scleroderma. Because rheumatoid disorders involve inflammation of the body's connective tissue, patients with Sjogren's syndrome are likely to have some kind of connective tissue disorder and suffer
from accompanying pain and discomfort.

Like other autoimmune disorders, the Sjogren's patient's immune system
turns on itself and attacks healthy tissue. In most cases of this syndrome, the
mucous membranes around the eyes and mouth that secrete moisture are usually the initial glands attacked by the immune system. Sjogren's syndrome is
believed to be caused by the overproduction of lymphocytes, which damage
these moisture-producing glands and thus diminishes the production of tears
and saliva by these glands. This can cause patients to have difficulty swallowing, in addition to light-sensitive eyes and ulcers on the eyes' corneas.
Healthy tissues in the lungs, kidneys, and liver can also be damaged.

Common symptoms of the disorder include dry eyes and mouth, dental
cavities, fatigue, low-grade fever, enlarged parotid glands (located behind
the jaw and in front of the ears), difficulty chewing or swallowing, nose-
bleeds, and bruising. Skin rashes, dry skin, and vaginal dryness are also
common symptoms. Of course, Sjogren's syndrome is difficult to diagnose
because these symptoms are common in other autoimmune disorders.

While patients of any age can develop Sjogren's syndrome, it was often
affects people over 40, and it is nine times more likely to affect women than
men. Patients who are suffering from any sort of rheumatic disease are also
more likely to develop this disorder, and it is believed to have a genetic
component, which means that heredity can play a factor.

As mentioned above, complications of Sjogren's syndrome include difficulty swallowing, dental cavities, and vision problems because dryness of
the eyes can lead to light sensitivity and corneal ulcers. In addition, scientists have linked heart problems in babies born to mothers who have been
diagnosed with Sjogren's syndrome.

Two common medications for dry mouth associated with this autoimmune disorder include pilocarpine (salagen) and cevimeline (evoxac). Both
drugs stimulate certain mouth glands to produce saliva, although common
side effects include excessive sweating, headaches, and nausea. Artificial
tears and eyes, which are available by both a prescription or over the
counter, can alleviate some of the discomfort associated with dry eyes.

GUILLAIN-BARRÈ SYNDROME

This disorder involves the peripheral nervous system, which is the division of the nervous system that consists of the cranial nerves (the brain's twelve pairs of nerves) and the spinal nerves (thirty-one pairs of nerves associated with the spinal cord). Also included in the peripheral nervous system is the autonomic nervous system (ANS), which controls the "automatic" or involuntary movements of the body's smooth muscles (found in the walls of tubes and hollow organs), cardiac muscles, and the glands. When a patient is diagnosed with Guillain-Barrè Syndrome (GBS), his body's immune system attacks portions of the peripheral nervous system. While a rare disorder (only affecting about one in 100,000), both men and women at any age are equally at risk for being diagnosed with the GBS.

Initial symptoms of GBS include weakness or tingling sensations in the legs, which then spread to the arms and upper body. If the disease progresses, certain muscles will completely weaken to the point of paralysis. GBS becomes life threatening if the paralysis spreads to the body's respiratory systems, in addition to interfering with the body's blood pressure and heart rate. If the progression gets severe, the patient will often be put on a respirator to assist with breathing, and will also be monitored closely for associated problems such as abnormal heart beat, infections, blood clots, and a sharp increase or decrease in blood pressure. Even though most GBS patients recover, significant weakness and exhaustion might persist. GBS can be fatal if the paralysis overtakes the heart, causing it to arrest.

Scientists and doctors do not yet know what causes GBS and why it strikes some patients rather than others, nor does anyone really understand all the specifics of the disease. In most cases, GBS will occur a few days or weeks following a patient's bout with a respiratory or gastrointestinal viral infection. On rare occasions, surgery or vaccinations can cause GBS (see "Guillain-Barrè Syndrome and the Flu Vaccine"). Development can occur over the course of hours or days, but in some cases it can take up to three or four weeks.

While much is not known about GBS, scientists and doctors know that in patients with this disorder, the immune system destroys the body's peripheral nerves by attacking a neuron's myelin sheath, which is composed of fatty material and electronically insulates neurons from one another. Without the protection of the myelin sheath, the neurons would short circuit and thus be unable to transmit electronic impulses. The myelin sheath surrounds the neuron's axons, which are single nerve fibers that carry impulses away from the cell body. Nerve fibers that carry impulses to the cell body are known as dendrites. The myelin sheath allows the nervous system to rapidly transmit signals to the muscles, brain, and sensory organs over long distances. More information on neurons can be found in the Nervous System volume of this series.

Guillain-Barrè Syndrome and the Flu Vaccine

In 1918, a horrific flu epidemic swept through the United States, killing approximately 40 million Americans. Doctors were on high alert to quell another outbreak, and the development of flu vaccines in later years was seen as the key to protecting future generations of Americans from succumbing to another epidemic. However, in 1976, health officials reported that over 532 people, 32 of whom died, came down with the Guillain-Barrè Syndrome after receiving a certain type of influenza inoculation called the swine flu vaccine. Even though the flu is blamed for about 20,000 deaths annually in the United States (mostly among the elderly), the CDC advises patients with a family history of Guillain-Barré not to get any flu vaccine, for fear that the predisposition for an adverse reaction could have a genetic component.

If the myelin sheath is damaged or destroyed, neurons cannot communicate and transmit signals normally. Therefore, muscles begin to behave abnormally and are unable to respond to the instructions coming from the brain. The brain's ability to receive information from sensory organs, such as the eyes, nose, and tongue is diminished, making it hard for the patient to feel sensations such as heat and pain. Not only is it hard for any sensory-related information to be transmitted, but sometimes inappropriate information is sent out through the body, resulting in tingling or painful sensations. Weakness and tingling sensations will often initially appear in the hands and feet and then move up throughout the body. This is because signals to the arms and legs must travel the longest through the body's nerve network, making them the most susceptible to interruption.

If a patient has a viral or bacterial infection and then develops GBS, some scientists and doctors believe that it is possible that the virus has caused cells in the nervous system to change so that the immune system treats them as foreign, and thus tries to destroy them. Another possibility is that the virus weakens the immune system, rendering it less discriminating about what cells it recognizes as normal and foreign. This could cause certain kinds of lymphocytes and macrophages to attack the myelin. For example, T lymphocytes that are sensitized will cooperate with B lymphocytes to make antibodies that react to certain components of the myelin sheath, and then work to destroy the sheath. These are some main areas of research into the workings and machinations of GBS that scientists are undertaking to better understand this autoimmune disorder.

Like many other autoimmune diseases, GBS has similar symptoms to other disorders of the immune system, so doctors make a thorough examination prior to deciding on a diagnosis. Doctors will look to see if the symptoms form a particular pattern that distinguishes GBS from other autoimmune disorders. For instance, GBS symptoms usually appear on both sides of the body. In addition, symptoms will come on quickly; muscle weakness will progress over days rather than months. Because GBS impacts nerve transmission, tests that analyze the strength of reactions can also help doctors determine if a patient has GBS. A spinal tap test can also reveal clues about GBS. By inserting a needle into a patient's lower back, doctors can take a sample of cerebrospinal fluid (which surrounds the brain and spinal cord) from the spinal column. If a patient has GBS, then the fluid will contain more proteins than usual.

Because there is currently no treatment for GBS, doctors rely on giving patients therapies that target the symptoms, which help to speed recovery in most patients. Two current treatments are called *plasmapheresis* and *high-dose immunoglobulin therapy*. Both of these treatments are effective, although the immunoglobulin therapy is considered easier to administer to patients.

When a patient is given plasmapheresis, whole blood is removed from the body, and then the plasma portion of the blood is separated from the red and white blood cells. Without the plasma, the blood cells are reintroduced into the body. Almost immediately, the body works to replace the plasma. Scientists and doctors do not exactly know why this process works. One theory is that the plasma in the blood might contain substances from the immune system that are harmful to the neurons' myelin.

High-dose immunoglobulin therapy involves intravenous injections of certain proteins from healthy donors. In small doses, these proteins are utilized by the immune system to attack invading or foreign pathogens. High doses of these donated immunoglobulins injected into GBS patients appear to lessen the immune system's assault on the nervous system.

The most critical element of treating a GBS patient involves keeping the body functioning while the nervous system is going through recovery. If a patient can't breathe on his own or is having heart problems, doctors will often use a respirator, a heart monitor, or other machine to help the body function close to normal. GBS patients are often required to stay in the hospital—usually in the intensive care ward—throughout recovery so doctors can treat the complications that accompany paralysis, such as pneumonia or bed sores.

CROHN'S DISEASE

This autoimmune disorder is considered chronic, and causes inflammation along the digestive or gastrointestinal (GI) tract, which extends from the mouth to the anus. In most cases of Crohn's disease, however, inflammation only affects the small intestine and/or colon. The disease is named after Dr. Burrill B. Crohn (1884–1983), who researched and reported on the condition in 1932. While considered an autoimmune disease, Crohn's and a related disease called *ulcerative colitis*, are also two primary types of inflammatory bowel diseases.

It is often difficult for doctors to distinguish between Crohn's and ulcerative colitis because the symptoms are very similar. In fact, according to the Crohn's and Colitis Foundation of America, doctors cannot determine whether approximately 10 percent of colitis cases are ulcerative colitis or Crohn's disease, so those cases are called *indeterminate colitis.*

Despite their minor differences, Crohn's and colitis are both autoimmune disorders and are characterized by abnormal behavior by the body's immune systems. When food, bacteria, and pathogens enter the intestine, the immune system mistakes these as dangerous substances and goes into attack mode. While in this state of attack, white blood cells are sent to the lining of the intestine, which causes chronic inflammation. These white blood cells are linked to producing harmful substances that eventually cause ulcerations and injury to the bowels.

Crohn's disease can affect the end of the small intestine and the beginning of the large intestine. In addition, it may involve any part of the GI tract. If a patient has ulcerative colitis, only the colon is involved. And while all layers of the intestine might be impacted by Crohn's disease, only the top layers of the colon are affected.

These disorders have been the focus of considerable research in recent years, but scientists and doctors still do not yet know exactly what causes these conditions. Some studies implicate a multitude of factors, including a genetic predisposition to irritable bowel–related disorders, a compromised immune system, and environmental factors. These environmental antigens may prompt the inflammation that is one of the primary symptoms of Crohn's and colitis diseases. Once the immune system begins producing this inflamed response, it does not know how to stop or reverse the response.

The primary symptoms of both Crohn's and colitis is persistent diarrhea, abdominal pain or cramps, fever, and in some cases, rectal bleeding. It is also common for the patient to experience a lack of appetite and subsequent weight loss. Fatigue and exhaustion are also common symptoms. In some patients, the tissue around the anus might develop slight tears, which can lead to severe pain and bleeding. These are chronic diseases, which means the symptoms cycle through periods of severity, but also go through mild periods. Active periods are considered flares and this is when symptoms are at their worst. These flares are then followed by mild periods when the disease appears to be in remission because the symptoms disappear or diminish to a significant degree.

Researchers estimate that approximately 1 million Americans suffer from either Crohn's disease or ulcerative colitis, and males and females appear to be affected equally. It can affect people of all ages, but the disorders have been found to mainly affect adolescents and young adults between the ages of 15 and 35. Scientists have estimated that approximately 10 percent of patients are under the age of 18.

As stated earlier, it appears as if genetics plays a role in getting this disorder. In fact, studies have shown that between 20 and 25 percent of patients have a close relative with either Crohn's disease or ulcerative colitis. If someone has a relative with either of these disorders, his risk of getting one of these diseases is ten times greater than someone without the family history. The risk rate climbs to thirty times greater if someone's brother or sister has either disorder.

Race and ethnicity appear to be a factor in these autoimmune disorders. Studies have shown that American Jewish people of European descent are four to five times more likely to develop Crohn's or ulcerative colitis in comparison to the general population. It also appears that these diseases primarily affect white people; the rate among the white population is 149 per 100,000 cases. In terms of environment, which is believed to play a role in both of these disorders, more cases are reported in urban areas as compared to rural, and more often in northern rather than southern climates.

While there is currently no cure for Crohn's disease or colitis, there are numerous medications and treatments available to control the body's inflammatory response. By controlling the inflammation, the tissues located in the intestines are able to heal. In addition, the fever, diarrhea, and abdominal pain symptoms are also alleviated with the decrease in inflammation. After these symptoms are controlled, doctors can identify treatments needed to control the frequency of flares in an effort to keep the disease in remission. Examples of Crohn's disease medications include aminosalicylates, corticosteroids, immune modifiers, and antibiotics.

The Influence of Nutrition on the Immune Response

As the previous chapters have indicated, the human lymphatic system, and the immune response that it controls, is an amazing network of interacting cells and chemical signals. As was described in Chapter 4, the immune system is active before a person even enters into the world, and for the majority of the population it functions correctly throughout their entire lifespan. The immune system not only remembers past diseases, but may be preprogrammed through the use of vaccinations. It has the ability to rapidly adapt as new infectious threats to the body emerge. It is not a static system, but a continuous dynamic response to the hostile organisms in the environment. While the lymphatic system and the immune response are responsible for maintaining health, they are also influenced by the same factors that affect other organ systems, namely an adequate supply of nutrients, including key vitamins and minerals.

NUTRITION AND THE IMMUNE SYSTEM

As is the case with every other organ system in the human body, nutrition plays a central role in the health of the lymphatic system and its ability to conduct an immune response. The primary problem with human nutrition is the conflicting information on the benefits of certain nutrients in preventing disease and infection. It is important to note that scientific studies regarding human nutrition, and the complex interactions of nutrients with the biochemical pathways in the body, are constantly in progress. However, the majority of nutritionists and medical professionals will agree that a well-balanced diet, consisting of an emphasis on fruits and vegetables, is essential to proper immune system function. The purpose of this section is to outline some of the major nutrients that are specifically required by the body in order to enhance the immune response. A more detailed description of nutrition is provided at the start of the Digestive System volume of this series.

VITAMINS

Vitamin A belongs to a class of chemicals called the *retinoids*. The retinoids have a variety of functions in the body, including the maintenance of vision systems and the growth of bones. With regards to the immune system, vitamin A helps in the development of the epithelial linings of the mucous membranes. These mucous membranes, such as the ones found in the nasal cavity and respiratory systems, act as the first line of defense against invading pathogens. Without vitamin A in the diet, these cells do not develop correctly, allowing an easy access for bacteria and viruses. A number of studies have shown that individuals with adequate intakes of vitamin A have less a reduced rate of infection and a reduction in the severity of disease symptoms. One of the retinoids, beta-carotene, is an **antioxidant** (discussed with vitamin C later in this section) and may protect against certain forms of cancer.

Vitamin A is found in milk and milk products, eggs, and liver. Beta-carotene is a plant retinoid found in most leafy green vegetables. Although adequate amounts of vitamin A are necessary for optimum health, vitamin A can be toxic in high doses (see "Are Vitamins Toxic?"). Toxicities are mostly associated with the use of vitamin supplements, so individuals should consult with their physician before taking high-dosage vitamin supplements.

Vitamin E is another of the antioxidant vitamins. Vitamin E belongs to a class of chemicals called the *tocopherols*. As is the case with vitamin A, vitamin E is associated with the health of the respiratory system. Among its many functions, vitamin E acts as an antioxidant for the respiratory system, protecting the lung tissues against damage and thus strengthening this first line of defense.

Vitamin E deficiencies are very rare, but are characterized by anemia, weakness, and problems with circulation. Vitamin E is found naturally in plant oils, liver, eggs, and leafy green vegetables. Vitamin E frequently works in cooperation with selenium (see following section, "Minerals").

Vitamin B_6, also known as pyridoxal, pyridoxine, and pyridoxamine, has an indirect association with the immune system. Vitamin B_6 is a precursor for an important enzyme that is involved in the amino acid metabolism. Amino acids are not only the building blocks of proteins (see the Digestive

Are Vitamins Toxic?

Although some of us may regard vitamins as being the miracle cure for many of life's ailments, the reality is that in high doses many vitamins may have toxic effects. This is primarily due to the body's relationship with vitamins. The body regards vitamins as trace nutrients, meaning that they are only required in small quantities daily. When exposed to high levels, especially over a period of time, the body attempts to store these nutrients in tissues. This can cause problems as metabolic pathways become overloaded. Some nutrients are especially prone to overdoses:

- Vitamin A can lead to improper bone development in children and problems with the liver and the spleen.

- Doses of over 1000 milligram a day of vitamin C can cause urinary tract problems, insomnia, and diarrhea.

- High doses of iron can cause hemochromatosis, which damages the tissues of the liver.

- Excessive levels of zinc can cause flulike symptoms (vomiting, diarrhea, fever).

- Selenium can be toxic in doses as little as 1 milligram and can cause nervous system problems.

It is always best to consult with a physician before beginning a new vitamin or mineral supplement.

System volume of this series), but are also used as the starting materials for a large number of important biochemicals in the body. Some of these components play roles in the health of the primary and secondary immune organs.

Studies have indicated that people who are vitamin B_6 deficient frequently have a lower ability to combat infectious diseases. Because vitamin B_6 is easily degraded by the use of alcohol, people who regularly use alcohol are especially susceptible to being vitamin B_6 deficient. Symptoms of a vitamin B_6 deficiency include cheilosis (cracks at the corners of the mouth), depression, fatigue, and anemia. Natural sources of vitamin B_6 include leafy green vegetables such as spinach, whole grains, and fruits. Heat destroys the vitamin B_6 content of foods.

Folate is an important vitamin in a wide range of body functions. Also known as pteroylglutamic acid (PGA), this vitamin is actively involved in the synthesis of DNA and is especially important in rapidly dividing tissues. Following an infection, the body frequently needs to quickly replace epithelial tissues (in the case of respiratory illnesses) or skin tissue (in the case of diseases such as chickenpox). Individuals who are deficient in folate may not be able to rapidly restore these physical barriers, leaving them open to secondary infections.

Folate is absorbed into the body in an inactive form, and must be activated by vitamin B_{12}. Folate is found in liver, green leafy vegetables, beans, and fortified grains and cereals. It is easily destroyed by heat and exposure to oxygen. Deficiencies in folate are often associated with a dietary deficiency in vitamin B_{12}. Vitamin B_{12} is contained primarily in the proteins of animal products. It requires a special transporter, called an intrinsic factor, for it to be absorbed by the GI tract. If the GI tract does not produce this intrinsic factor, possibly due to an injury or disease, then vitamin B_{12} will not be absorbed, potentially resulting in a folate deficiency. Vegetarians are especially likely to have vitamin B_{12} deficiencies due to the nature of their diet.

In many ways, vitamin C is one of the least understood and most misused vitamins with regards to immune function. Vitamin C, also called ascorbic acid, is, like beta-carotene, an **antioxidant**. In the body, vitamin C protects many of the other vitamins from being damaged or destroyed by oxidation reactions. It also serves to protect the body's tissues from attack by **free radicals** or other damaging agents. During an immune response, large amounts of free radicals are produced, and thus vitamin C helps prevent the immune response from inadvertently damaging the tissues of the body. In each of these cases, the vitamin C molecule is destroyed in the process.

Like the other vitamins, a vitamin C deficiency can occur for a variety of reasons. The most common of which is of course a dietary deficiency. Vitamin C is found in fresh fruits, namely citrus fruits (oranges and grapefruits, for example), but also in dark green vegetables. Diets that are poor in these food products result in a suppressed immune system and increased susceptibility to infections. Stress also reduces vitamin C levels in the body, because the stress response produces free radicals, which in turn destroy vitamin C. This is one of the reasons why people in high-stress environments frequently have higher incidence of infections.

Many people falsely believe that vitamin C can cure the common cold. A number of studies have indicated that vitamin C does not significantly reduce the duration of the cold, but it can help reduce the effects of certain symptoms. A diet that contains adequate amounts of vitamin C can help prevent infections, such as colds, by aiding the immune response. Vitamin C has also been demonstrated to lower the risk of heart disease and certain types of cancer. Most nutritionists recommend a diet that contains around 400 milligrams per day of vitamin C.

MINERALS

The mineral iron is well known for its association with red blood cells and the circulatory system. Iron is a key component of hemoglobin, the primary oxygen transport protein of humans. Deficiencies in iron cause a decrease in the number of red blood cells (anemia) and reduced metabolic efficiency by the body. People with anemia and iron deficiencies usually also have a suppressed immune system, making them more susceptible to colds and other illnesses.

Iron is found in meats and poultry and in low quantities in leafy green vegetables. Unfortunately, iron is not readily absorbed by the intestinal tract (see the Digestive System volume of this series) and thus must be regularly supplied in the diet. Supplements may be used to help reduce this problem, although high-dosage supplements can cause problems (see "Are Vitamins Toxic?"). Vitamin C helps aid iron absorption, and thus citrus fruits should be eaten with iron-rich foods.

Zinc is an all-purpose mineral that is involved in over 300 metabolic pathways in the human body. In the immune system, zinc is important in the development of both white blood cells and the lymphocytes, and is needed to maintain the health of the thymus. People with zinc deficiencies frequently have a suppressed immune system, which leaves them susceptible to frequent illnesses.

Zinc deficiencies are caused by diets that are lacking in zinc-rich foods, such as meats, fish, poultry, and whole grains. Children are especially susceptible to zinc deficiencies, because their food consumption is frequently not adequate to supply enough zinc-rich food for their rapidly growing tissues and organs. However, care should be taken before using a zinc supplement, for both adults and children, because the threshold between adequate zinc in the body and toxicity is very narrow.

Selenium is an antioxidant mineral that works very closely with vitamin E. The exact relationship between selenium and the immune response is not clearly understood, but people who are selenium deficient have a higher susceptibility to viruses and certain forms of cancer. Selenium is a mineral that is found in the soil, so deficiencies are mostly associated with crops grown in selenium-poor soil. Selenium deficiencies are rare in North America, but common in areas such as China. As was the case with zinc, selenium can be toxic at doses as small as 1 milligram.

Acronyms

AIDS	Acquired immunodeficiency syndrome
APC	Antigen-presenting cell
BD	Behçet's disease
CAT	Computerized axial tomography
CFS	Chronic Fatigue Syndrome
DNA	Deoxyribonucleic acid
DTP	Diphtheria, tetanus, pertussis
FDC	Follicular dendritic cell
GAS	Group A Streptococcus
GI	Gastrointestinal
HAART	Highly active antiretroviral therapy
ICM	Inner cell mass
IL	Interleukin
MDS	Myelodysplastic syndromes
MHC	Major histocompatibility complex
MMR	Measles, mumps, rubella
MRI	Magnetic resonance imaging
NIH	National Institutes of Health
NK	Natural killer cells
PGA	Pteroylglutamic acid
RBC	Red blood cells
RNA	Ribonucleic acid
SARS	Sudden acute respiratory syndrome
WBC	White blood cell
WHO	World Health Organization

Glossary

ABO group The name of the genetic system that determines human blood groups. Named for the presence of A and B carbohydrates on the surface of the cell, or the absence of the carbohydrates in the case of the O group. This system uses four possible combinations: A, B, AB, or O.

Acquired immunity A type of immunity that is not the result of genetic inheritance, but rather due to the exposure to some antigen and the resulting response by the immune system.

Adrenal gland The hormone-releasing gland located above the kidneys.

Afferent vessels A form of vessel that brings fluid towards an organ or lymph node.

Agglutination The process by which objects come together and adhere to one another as if held together by glue. In the immune system, this "glue" is usually an antibody.

Allele A variation of a gene that encodes for a specific trait. It is usually due to minor variations in the DNA at the molecular level.

Allergen Any substance that invokes an allergic reaction in the body.

Allergy An exaggerated response to an environmental allergen that is not a pathogen; this usually involves the IgE class of antibodies.

Anaphylaxis An increased susceptibility to a foreign protein resulting from a previous exposure; may cause a serious condition called anaphylactic shock.

Anemia A reduction in the number of red blood cells in the body. As a result, the number of hemoglobin molecules is insufficient to carry oxygen to the tissues of the body. This may result in tissue color changes, weakness, and increased susceptibility to disease.

Aneurysms A weakening of the wall of a blood vessel; this causes a dilation of the blood vessel and a disruption of normal blood flow.

Antigen Any substance that initiates the specific immune response in the body.

Antioxidant A chemical that inhibits oxidation reactions by donating electrons to another molecule to protect it against oxidation.

Appendectomy The surgical procedure that is used to remove an inflamed, diseased, or ruptured appendix.

Appendicitis An inflammation of the appendix. This is usually caused by an infection of the appendix and results in fever, pain, and loss of appetite.

Asymptomatic A person who does not show the symptoms of a disease. Many people may be asymptomatic in the early stages of a disease.

Autoimmune disease This occurs when the immune system incorrectly identifies the tissues of the body as foreign material, and begins an immune response against the cells or tissue. Lupus and forms of diabetes may be caused by an autoimmune response.

Benign A condition that does not spread or get progressively worse and responds to treatment.

Biomolecule A general classification for any of the four groups of organic molecules that are used in the building of cells—proteins, carbohydrates, lipids, and nucleic acids.

Biopsy A medical procedure by which a small sample of tissue is removed for study; this is usually done using a small needle. A pathologist then examines the tissue under a microscope for signs of a specific disease, such as cancer.

Blastocyst An early embryonic form comprised of a round, hollow cavity filled with fluid that is surrounded by a single layer of cells.

Blastula The hollow ball of cells that forms during the first week of human embryonic development.

Bone marrow The site in the body where the cells of the lymphatic system originate.

Carcinogens Cancer-causing agents or substances.

Carcinoid tumor A form of tumor that originates in the intestinal cells of the body.

Carcinoma A form of cancer that originates in the epithelial cells of the body and then spreads, or metastasizes, into other organs of the body.

Cell wall In fungi, bacteria, plants, and some protistans, this is a semi-rigid covering of the cell located externally to the cell membrane. It is made of complex carbohydrates, although the composition varies between groups.

Chemotaxis The reaction of mobile cells to a chemical gradient; the cells may move either towards or away from the gradient depending on the nature of the chemical being used.

Chemotherapy Often used to describe the treatment of cancer using chemicals, the term can also be applied to the treatment of microorganisms using antibiotics or other chemical mechanisms.

Clinical trials A form of scientific medical study that uses volunteers as test subjects in an attempt to assess the effectiveness of a new procedure or drug.

Clots The result of blood coagulation. Coagulation is based on the interaction of proteins present in the plasma of the blood.

Coagulation The transition of a liquid to a semi-solid gel with varying degrees of consistency.

Complement fixation The process by which complement factors bind to either antibodies or cell surfaces during the immune response.

Computerized axial tomography (CAT) A diagnostic tool in which a radioactive compound is given to a patient, and the absorption of the compound by different tissue types is recorded by a detector. It is frequently used in the study of the central nervous system.

Cortisol A steroid produced by the adrenal cortex that is responsible for maintaining blood pressure, as well as metabolizing carbohydrates.

Cytokine A chemical signal of the immune system.

Diphtheria A lymphatic disease where the presence of bacteria leads to the production of a toxin that infects the mucous membrane of the throat and other respiratory passages. Patients with this disease experience breathing difficulties, as well as a high fever.

Dysentery A disease characterized by an inflammation of the intestines; the result is usually diarrhea, dehydration, and blood in the stool.

Edema A swelling beneath the skin's surface due to fluid build up in cellular tissue.

Efferent vessels A vessel of the lymphatic or circulatory systems that carries fluid away from an organ or lymph node.

Elephantitis A disease that results from the blocking of the lymphatic vessels, causing fluid to build up in the tissues. The result can be enlargements of the extremities, usually the legs.

Epinephrine Also called adrenaline, this hormone is produced in response to stress.

Erythropoietin A hormone produced by special cells in the kidney in response to oxygen deprivation in tissues of the body. The result is an increase in red blood cell production in the bone marrow.

Extranodal A term used to describe conditions that are not associated with the lymph nodes.

Extrinsic factors Another term used for vitamin B_{12} in the diet.

Fever The abnormal increase of body temperature that is often the symptom of an illness.

Folliculitis Any inflammation of the follicles of the body; usually used in reference to the hair follicles.

Free radical A highly unstable molecule that contains at least one unpaired electron. These molecules will rapidly react with organic material, frequently causing damage.

Fungi One of the six recognized kingdoms of living organisms. Fungi are a diverse group of nonphotosynthetic organisms that are frequently called the *decomposers*; some members of this group are pathogenic.

Genes The fundamental unit of inheritance. Genes contain the instructions for forming proteins, which in turn are responsible for all of the physical, physiological, and metabolic traits of an organism.

Genetic immunity A form of immunity to a pathogen that is inherited.

Glucocorticoids Hormones that are produced by the adrenal cortex. These hormones manage metabolism of carbohydrates, proteins, and fat, and also have anti-inflammatory characteristics.

Glycoproteins A type of protein that has a complex carbohydrate attached to it. These are typically located within the cell membrane and are involved in cell signaling.

Graft-versus-host disease A disorder that can result from a tissue transplant, which occurs when the donor's transplanted tissue attacks the cells of the recipient of the transplant.

Hemolysis (haemolysis) The breakdown of the membranes of red blood cells and the subsequent release of hemoglobin from the cell.

Herpes An inflammation of the skin caused by the herpes simplex virus.

Hydrophilic A "water-loving" molecule, meaning that it will interact freely with water.

Hydrophobic A "water-hating" molecule, meaning that it will not freely interact with water.

Hypoxia The reduction of the oxygen supply to cells, tissues, or organs to a level that is below physiological minimums.

Immunity The ability of an organism not to be affected by a given disease or pathogen.

Immunochemistry An area of the chemical and biological sciences that studies the chemical nature, signals, and properties of the cells and organs of the immune system.

Immunoglobins Another name for antibodies; these are specific proteins produced by B cells in response to a pathogen.

Impetigo A skin disorder caused by a staphylococcus bacteria.

Insulin A hormone produced by the pancreas that signals the cells of the body to increase glucose absorption, reducing blood glucose levels.

Intrinsic factors A protein released by the gastrointestinal tract that aids in the absorption of vitamin B_{12}.

Isotope Two atoms belonging to the same element but that have a different atomic weight; the difference is due to the number of neutrons in the nucleus. Some isotopes are radioactive compounds.

Larva A free-living juvenile form of some animals that usually differs morphologically from the adult. Many insects possess distinct larval stages.

Lipoprotein A type of protein that has a lipid attached to it. These frequently are used in the body to transport fats, cholesterol, and fat-soluble vitamins in the lymphatic and circulatory system.

Lymph The name for the fluid of the lymphatic system. It is a colorless fluid that is derived from the interstitial fluids of the body.

Lymphangitis The inflammation of a lymphatic vessel, usually as a result of a bacterial infection.

Lysozyme An enzyme that is present in tears and perspiration that functions as a nonspecific defense against bacteria.

Magnetic resonance imaging (MRI) A diagnostic tool based on the fact that hydrogen atoms resonate or vibrate at distinct frequencies when bombarded by energy from a high-power magnet. This tool produces a three-dimensional image of the tissues being studied, and is very useful in examining minor variances in body chemistry.

Malignant A condition that becomes progressively worse or more pronounced over time, and which may lead to death.

Medullary cords Within a lymph node, these are areas of dense lymphatic tissue.

Megaloblasts Oversized red blood cells often found in people with anemia or other vitamin-deficiency disorders.

Meningitis An inflammation of the meninges of the nervous sytem.

Metastasis The movement of a disease, such as cancer, between two unconnected organs.

Monoclonal antibodies Antibodies that are produced in the lab from a single cell. They are very useful in scientific research to study the effects of a specific antigen or disease.

Monocytes The largest of the white blood cells.

Myelodysplasia When the cells of the bone marrow form incorrectly, or are defective.

Neutrophils A type of white blood cell that recognizes and destroys antigens in the blood stream.

Nicotine An addictive drug with poisonous properties that is derived from tobacco plants.

Normoblast The cells of the bone marrow that are responsible for the formation of the red blood cells.

Notochord A flexible, rod-like structure in the embryos of higher vertebrates, from which the spinal column develops.

Nuclear medicine An area of the medical sciences that uses radioactive isotopes for the diagnosis and treatment of disease.

Opportunistic infection A form of infection that is usually rare in the general population, but occurs frequently in individuals with compromised immune systems.

Opsonization The modification of a bacteria so that it is more easily recognized by the immune system, resulting in an increase in phagocytosis by macrophages.

Osmosis The movement of fluid from an area of high concentration to an area of low concentration.

Pedigree In genetic analysis, the tracing of the ancestral lineage of an individual; the process often utilizes a graphic representation, also referred to as a "family tree."

Phagocytic cell A type of cell that engulfs external particles, food, or organisms into its cytoplasm; the enclosed material may then be destroyed by digestive enzymes.

Plasma The fluid that comprises the majority of the blood.

Platelets A component of the clotting response. Platelets form in the bone marrow.

Pollen The reproductive structure of plants that carries the male sperm. Pollen frequently is an allergen.

Polygenic A disease or trait that is due to the effects of several genes. Human eye color and skin color are examples of polygenic traits.

Precipitate The portion of a chemical reaction that does not remain soluble in the original liquid and falls out of the solution.

Prostaglandin A type of hormone group that controls various physiological functions, including blood pressure, smooth muscle contractions, and inflammatory reactions.

Protease A class of enzyme that is involved in the breakdown of proteins into amino acids.

Protistans One of the six recognized kingdoms of life. Members of this diverse kingdom are unicellular organisms with a nucleus and membrane bound organelles.

Radiation therapy The treatment of a disease or condition using concentrated bursts of x-rays or other radiation.

Reticulocyte Immature red blood cells; these are usually found in the bone marrow.

Rh factor An antigen that is found on the surface of blood cells; it is an independent factor of the ABO group. The plus designation indicates the presence of the antigen, the negative indicates the absence.

Rhinitis An inflammation of the mucous membranes of the nasal passage.

Scientific method The basic premise of modern science; it involves the collection of data, the formulation of a hypothesis, experimentation to test the hypothesis, and an assessment of the results.

Serum The clear portion of blood, without blood cells and platelets.

Sickle cell anemia A human genetic disease caused by a mutation in which one amino acid in the hemoglobin protein is incorrect.

Spermicide Any chemical that kills sperm cells; usually used to increase the effectiveness of contraceptives.

Synovium The inner layer of a synovial joint that is responsible for the secretion of the lubricating synovial fluid.

Syphilis A sexually transmitted disease caused by spiral bacteria *Treponema pallidum*.

Tetanus A disease that is caused by toxins from *Clostridium* bacteria; it is characterized by involuntary contractions of the muscles. The disease is sometimes called *lockjaw*.

Thrombophlebitis Following the obstruction of a vein (called a *thrombus*), this is the condition in which the vein becomes inflamed.

Thymus In humans, an organ just behind the heart in which the T cells of the body mature; this organ gets steadily smaller as a person ages.

Tonsillectomy A medical procedure that involves the removal of the tonsils, usually as the result of repeated infections.

Tonsils The name given to the lymphatic tissue found at the back of the oral cavity.

Toxoid The toxin produced by a bacteria which has been detoxified, but still retains its antigen characteristics. Toxoids are useful in the generation of immunizations.

Triglycerides The most common form of lipid in the human body. This molecule has three fatty acid chains covalently bound to a single glycerol molecule. Its primary role in the body is energy storage.

Tumor A mass of cells that results from abnormal cell division. Tumors may be either malignant or benign.

Ultrasound A medical procedure in which sound waves are applied to the body. The reflections of these waves are detected by a specialized instrument and provide a view of the internal structures of the body. This procedure is useful in examining tissues that normally would not be detectable by x-ray examination.

Uvula The finger-like projection in the rear of the oral cavity. Its purpose is to seal off the nasal cavity during swallowing.

Vascular A structure that is involved in the transportation of fluid.

Vasodilation The relaxation of the muscles surrounding the vascular tissue; this increases the diameter of the vessel and reduces pressure.

Vena cava One of two large veins that returns deoxygenated blood from the tissues of the body into the right atrium of the heart.

Virus A nonliving infectious agent that is characterized as having a protein covering and either DNA or RNA as its genetic material; some viruses may also have a lipid covering. Viruses are completely dependent on cells for reproduction.

Zygote The name given to the cell that is formed by the fusion of the egg and sperm cell during fertilization.

Organizations and Web Sites

American Autoimmune Related Diseases Association
22100 Gratiot Avenue
East Detroit, MI 48021
Phone: (586) 776-3900
Literature requests: (800) 598-4668
www.aarda.org/

The homepage of this organization, which provides news and links to other sites for information on the major autoimmune diseases.

American Cancer Society
1599 Clifton Road
Atlanta, GA 30329
Phone: (800) ACS-2345
www.cancer.org

A site for all aspects of cancer prevention, diagnosis, and treatment, including cancers of the lymphatic system.

The Anatomy of the Immune System
www-micro.msb.le.ac.uk/MBChB/2b.html

Designed by the Department of Microbiology and Immunology at the University of Leicester, this brief but informative site contains an overview of the anatomy of the major organs in the lymphatic system. It also includes several informative schematic diagrams of the system.

Autoimmune Diseases Online
www.autoimmune-disease.com

Primarily contains links to organizations associated with autoimmune diseases.

The Body
www.thebody.com

An educational Web site on HIV/AIDS. Includes information on prevention, treatment, support groups, current research, and information on improving the quality of life for HIV/AIDS patients.

"How Your Immune System Works"
http://www.howstuffworks.com/immune-system.htm

This site contains nontechnical information on the structure and operation of the immune system. It is an ideal reference site for individuals without a strong science background.

Leukemia and Lymphoma Society
1311 Mamaroneck Avenue
White Plains, NY 10605
Phone: (914) 949-5213
www.leukemia.org

Contains information on both leukemia and lymphoma for patients and medical practitioners. Also includes recent news on advances in research.

Lymphatic Research Foundation
39 Pool Drive
Roslyn, NY 11576
Phone: (516) 625-9675
www.lymphaticresearch.org/

A private foundation that is dedicated to researching cures for a variety of lymphatic disorders.

Understanding the Immune System
press2.nci.nih.gov/sciencebehind/immune/immune01.htm

Designed by the National Cancer Institute, this site contains detailed information on the operation of the lymphatic system and the immune response.

Bibliography

Allergy Report, www.theallergyreport.org/reportindex.html.

American Academy of Allergy, Asthma and Immunology, www.aaaai.org.

American Academy of Family Physicians, www.familydoctor.org.

American Academy of Otolaryngology, www.entnet.org.

American Medical Association, American Academy of Pediatrics, Medical Library, www.medem.com.

Aplastic Anemia and MDS International Foundation, Inc., www.aamds.org

Asimov, Isaac. *Asimov's Chronology of Science and Discovery*. New York: Harper-Collins, 1994.

———. *Asimov's New Guide to Science*. New York: Basic Books, Inc., 1984.

Banchereau, Jacques. "The Long Arm of the Immune System." *Scientific American* 287, no. 5 (November 2002): 52–59.

Bibel, Debra Jan. *Milestones in Immunology: A Historical Exploration*. Madison, WI: Science Tech Publishers, 1988.

Canadian Hemophilia Society, www.hemophilioa.ca.

Centers for Disease Control and Prevention, www.cdc.gov.

Chandra, R. K. "Nutrition and the Immune System from Birth to Old Age." *European Journal of Clinical Nutrition* 56, Suppl. 3 (2002): S73–S76.

Crohn's and Colitis Foundation of America, www.ccfa.org.

Eales, Nellie B. "The History of the Lymphatic System, with Special Reference to the Hunter-Munro Controversy." *Journal of the History of Medicine and Allied Sciences* 29, no. 3 (1974): 280–294.

Healthlink: Medical College of Wisconsin, http://healthlink.mcw.edu.

Hellemans, Alexander, and Bryan H. Bunch. *The Timetables of Science: A Chronology of the Most Important People and Events in the History of Science*. New York: Simon & Schuster, 1988.

Hoffman-Goetz, Laurie, ed. *Exercise and Immune Function*. Boca Raton, FL: CRC Press, 1996.

Knight, Bernard. *Discovering the Human Body*. New York: Lippincott & Crowell Publishers, 1980.

Mayo Clinic, www.mayoclinic.com.

Melton, Lisa. "Subduing Suppressors." *Scientific American* 287, no. 6 (December 2002): 28–30.

Merck Manual of Medical Information, www.merck.com/mrkshared/mmanual_home2/home.jsp.

National Cancer Institute, www.cancer.gov.

National Hemophilia Foundation, www.hemophilia.org.

National Institute of Arthritis and Musculoskeletal and Skin Diseases, www.niams.nih.gov.

National Institutes of Allergy and Infectious Diseases, www.niaid.nih.gov.

National Institutes of Health, www.nih.gov.

National Lymphedema Network, www.lymphnet.org.

Nemours Foundation, www.kidshealth.org.

Nieman, David C., and Bente Klarlund Pedersen. *Nutrition and Exercise Immunology*. Boca Raton, FL: CRC Press, 2000.

Parham, Peter. *The Immune System*. New York: Garland Publishing, 2000.

Raven, Peter H., and George B. Johnson. *Biology*, 5th ed. Boston, MA: WCB/McGraw-Hill Publishers, 1999.

Sanders, Tina, and Valerie C. Scanlon. *Essentials of Anatomy and Physiology*, 3rd ed. Philadelphia: F.A. Davis Company, 1999.

Serafini, Anthony. *The Epic History of Biology*. Cambridge, MA: Perseus Publishing, 1993.

Sherwood, Lauralee. *Human Physiology: From Cells to Systems*, 4th ed. Pacific Grove, CA: Brooks/Cole, 2001.

Silverstein, Arthur M. *A History of Immunology*. New York: Academic Press, 1989.

Singer, Charles, and E. Ashworth Underwood. *A Short History of Medicine*, 2nd ed. New York: Oxford University Press, 1962.

Sompayrac, Lauren. *How the Immune System Works*. Malden, MA: Blackwell Science, 1999.

Sweeney, Lauren J. *Basic Concepts in Embryology: A Student's Survival Guide*. New York: McGraw-Hill Companies, 1998.

Symons, Alan. *Nobel Laureates: 1901–2000*. London: Polo Publishing, 2000.

Tortora, Gerard J., and Nicholas P. Anagnostakos. *Principles of Anatomy and Physiology*, 2nd ed. San Francisco: Canfield Press, 1978.

U.S. Department of Health and Human Services, The National Women's Health Information Center, www.4woman.gov.

Wainwright, Milton. *Miracle Cure: The Story of Penicillin and Golden Age of Antibiotics*. Cambridge, MA: Basil Blackwell, 1990.

Whitney, Eleanor N., and Sharon R. Rolfes. *Understanding Nutrition*, 8th ed. Belmont, CA: West/Wadsworth Publishers, 1999.

Windelspecht, M. "Gene Therapy." In *Magill's Medical Guide*, 3 rev. ed. Pasadena, CA: Salem Press, 2004.

———. *Groundbreaking Scientific Experiments, Inventions and Discoveries of the Seventeenth Century*. Westport, CT: Greenwood Publishing, 2002.

———. *Groundbreaking Scientific Experiments, Inventions and Discoveries of the Nineteenth Century*. Westport, CT: Greenwood Publishing, 2003.

Index

ABO group. *See* Blood types

Acquired immunity, 50–53

Acquired immunodeficiency syndrome (AIDS), 1, 18, 103–9, 130, 132; lymphoma and, 100–101; treatment, 106–8; vaccine, 108–9

Addison, Thomas, 70

Adenoid, 34–35

Adrenaline. *See* Epinephrine

Aging, and immune response, 55–56

AIDS. *See* Acquired immunodeficiency syndrome (AIDS)

Allergies, 125–32; contact dermatitis, 128–29; food, 130–31; insect, 132; latex, 129–30; rhinitis, 126–27

Anaphylaxis, 126, 131

Anemia, 26, 109–11, 120, 148; sickle-cell, 111

Aneurysms, 137

Anthrax, 71–72, 81

Antibiotics, 41; history of study, 83–85; resistance to, 85. *See also* names of specific antibiotics

Antibodies, 23–24, 34, 40–41, 46, 50, 54, 79, 134; and allergies, 125–26; and blood types, 28; classes of, 40; and complement proteins, 3; his-

tory of study, 79–81, 81–83; monoclonal, 83, 100; production of, 8–12, 52, 56, 120. *See also* B cells; Plasma cells

Antigen, definition of, 8

Antigen-presenting cell (APC), 14, 17–18

Antihistamines, 18, 127, 131

Anti-inflammatory drugs, 50

Antimicrobial proteins, 4–5

Antioxidants, 148, 150–51

Antitoxins. *See* Toxins

Appendicitis, 35

Appendix, 8, 35; history of study, 65

Aristotle, 65

Arthritis, 136; rheumatoid arthritis, 135

Ascorbic acid. *See* Vitamin C

Aselli, Gaspare, 66–67

Aspirin, 50

Asthma, 127–28. *See also* Allergies

Autoimmune response, 48, 132–34; diseases of, 8, 102, 135–45; medications, 135. *See also the names of specific diseases*

B cells, 8–12, 18, 42–44, 46, 56; antibody production of, 8–12; and au-

About the Authors

JULIE McDOWELL is an independent scholar and science journalist. She is the author of *The Nervous System and Sense Organs* in Greenwood's *Human Body Systems* series. She is also a former assistant editor for two science publications, *Today's Chemist at Work* and *Modern Drug Discovery*.

MICHAEL WINDELSPECHT is Assistant Professor of Biology at Appalachian State University. He is the author of two books in Greenwood's *Groundbreaking Scientific Experiments, Inventions, and Discoveries through the Ages* series, author of *The Digestive System* in the *Human Body System* series, and editor of the *Human Body Systems* series.